Food Microbiology
In Human Health and Disease

Food Microbiology
In Human Health and Disease

Editor

Nancy M. Khardori

Division of Infectious Diseases
Department of Internal Medicine
Eastern Virginia Medical School
Norfolk, VA, USA

CRC Press
Taylor & Francis Group
Boca Raton   London   New York

CRC Press is an imprint of the
Taylor & Francis Group, an **informa** business

A SCIENCE PUBLISHERS BOOK

CRC Press
Taylor & Francis Group
6000 Broken Sound Parkway NW, Suite 300
Boca Raton, FL 33487-2742

© 2016 by Taylor & Francis Group, LLC
CRC Press is an imprint of Taylor & Francis Group, an Informa business

No claim to original U.S. Government works

Printed on acid-free paper
Version Date: 20151103

International Standard Book Number-13: 978-1-4987-0878-4 (Hardback)

Library of Congress Cataloging-in-Publication Data

Names: Khardori, Nancy.
Title: Food microbiology : in human health and disease / [edited by] Nancy Khardori.
Description: Boca Raton : Taylor & Francis, 2016. | "A CRC title." | Includes bibliographical references and index.
Identifiers: LCCN 2015041126 | ISBN 9781498708784 (hardcover : alk. paper)
Subjects: LCSH: Foodborne diseases. | Food--Microbiology.
Classification: LCC QR201.F62 F68 2016 | DDC 615.9/54--dc23
LC record available at http://lccn.loc.gov/2015041126

Visit the Taylor & Francis Web site at
http://www.taylorandfrancis.com

and the CRC Press Web site at
http://www.crcpress.com

Preface

The world of microbes in and around us is vast and diverse. It is now well accepted that the human body contains more microbial cells than human cells. The well-studied host-associated bacterial flora is in the gastrointestinal tract which harbors more than tenfold microbial cells than the number of human cells in the entire body. The association of this "Natures Bioshield" with the human host is strong and symbiotic.

Food is the most significant factor involved in modulating the bacterial flora in the GI tract. Changes in eating habits and environmental patterns are major causes of worldwide increase in obesity and gut microflora dysbiosis. Recent studies have established a link between gut microbes, low grade inflammation, compromised immune system and metabolic disorders.

In spite of the defense provided by the resident flora, GI associated lymphoid system and secretory IgA; infections caused by pathogenic bacteria entering through the gastrointestinal tract are common.

Food-borne infectious diseases are responsible for significant morbidity and measurable mortality. Advances in food processing technology and medical care have not decreased the prevalence of food-borne infectious diseases. On the contrary increase in global travel and food trade has added a major dimension to the area of food-borne illnesses.

The use of antibiotics in animal feed has impacted the development of resistance in human pathogens that subsequently cause difficult to manage serious infections.

This book reviews the multiple facets of food microbiology as applied to human health and disease. It starts with a historical perspective and overview of food-borne infectious diseases, discusses in detail the recent food-borne infectious disease outbreaks in the United States and their investigation. Unless the sources of contamination and vehicles for transmission are understood, the control of food-borne infection is not possible. Chapter five details the sources and vehicles followed by molecular epidemiology of food-borne infection in Chapter six. The clinical approach to food-borne infectious diseases in the form of diagnosis, treatment and prevention is discussed in Chapter seven.

The Chapter on antibiotic use in animal feed addresses complex interaction between antibiotic use, animal microbial flora, human microbial flora and human disease. Successful enforcement strategies are already in place in the European Union. The recent modest steps by US FDA and USDA towards addressing this significant public health issue were long awaited and are a reason for hope. The most fascinating aspect of food microbiology is the role it plays in metabolic disorders, in particular the world wide epidemic of diabesity. This book ends with a Chapter on the role of prebiotics and probiotics in the management of obesity, an approach likely to prove much safer than the "obesity drugs". To quote Sir William Osler "knowledge makes possible cure or control of disease".

Nancy M. Khardori, MD, PhD
Division of Infectious Diseases
Department of Internal Medicine
Eastern Virginia Medical School
Norfolk, Virginia

Contents

CHAPTER 1

Foodborne Infectious Diseases—Historical Perspective and Overview

Catherine F. Derber[1] and *L. Beth Gadkowski*[1,*]

Introduction

Foodborne infections constitute a significant worldwide public health problem (Kuchenmüller et al. 2009). In the United States, foodborne illness affects 1 in 6 Americans annually, resulting in an estimated 48 million episodes of illness, 128,000 hospitalizations and 3000 deaths (Scallan et al. 2011a, Scallan et al. 2011b). Norovirus, nontyphoidal *Salmonella* species, *Clostridium perfringens* and *Campylobacter* account for the majority of these illnesses, with a growing number of illnesses attributed to lesser known emerging organisms (Scallan et al. 2011a, Scallan et al. 2011b). The disease results from ingesting food and food products that are contaminated with bacterial, viral, and parasitic microorganisms or chemicals at any point in the food production and distribution process. Transmission may occur from any foodstuff, with increasing recognition of fresh fruits and vegetables as the source of many disease outbreaks (Berger et al. 2010, Painter et al. 2013). Linking sporadic illness to a particular food is difficult since most organisms are transmitted by more than one food source. As a result, outbreaks may be the only way to implicate a food vehicle definitively (CDC 2013). Even so, about one third of the reported foodborne disease outbreaks in the US are of unknown etiology (CDC 2013). While largely preventable,

[1] Division of Infectious Diseases, Department of Internal Medicine, Eastern Virginia Medical School, Norfolk, VA, USA.
* Corresponding author: gadkowlb@evms.edu

foodborne illnesses remain common and constitute a substantial economic burden; recent data estimates foodborne illness in the US costs more than $15.6 billion dollars annually (http://www.ers.usda.gov/data-products/cost-estimates-of-foodborne-illnesses.aspx#.VDW27r4mUfy).

Historical Perspectives

Foodborne illnesses have long been a source of morbidity and mortality. It has been postulated that Alexander the Great was not poisoned, but instead died of typhoid in 323 BC (Oldach et al. 1998). Similar claims have been made regarding the role of typhoid in the failure of Jamestown, the first English colony in the Americas (Anderson 2011). "Enteric fever", as typhoid was called in the mid-nineteenth century, has been implicated in the death of US President William Henry Harrison, and likely contributed to that of Zachary Taylor (McHugh and Mackowiak 2014). While little concrete data exists to support these assertions, there is no denying the impact that foodborne illness has had on history.

The epidemiology and management of foodborne infections has evolved over many years in response to increasing population size and advancing technology. Foodborne infections have decreased in part to improved sanitation and storage of food, however the ability to travel and trade food products internationally has hailed the emergence of new sources and new modes of transmission (Keusch 2013).

The cholera epidemic in mid-nineteenth century London highlights one of the major developments in food safety. In the middle of nineteenth century, Dr. John Snow discovered that the transmission of cholera was associated with consumption of contaminated drinking water. Prior to that time, authorities believed cholera epidemics resulted from miasma, a toxic stench caused by rotting vegetation and animal matter found in human waste (Hempel 2013, Markel 2013). Indeed, the cholera epidemic in London became known as the "Great London Stink" (Keusch 2013).

Two seminal observations helped Snow conclude that cholera was spread by the fecal-oral route. The first involved a baby infected with cholera in the Soho district of London. The baby's mother washed dirty cloth diapers in a cesspool which shared a water supply with the community water pump (Newsom 2006). Snow proposed to remove the water pump handle and encouraged meticulous hand hygiene. The incidence of cholera subsequently dropped significantly (Markel 2013).

The second observation involved a large number of people infected with cholera whose main water supply through London's Lambeth, Southwark, or Vauxhall water companies all derived from the Thames River. Sewage from multiple communities fed into the Thames River. When the Lambeth water company moved upstream, the number of people infected with cholera decreased dramatically in proportion to people using the other companies (Hempel 2013). William Heath famously depicted society's early concerns about London's drinking water in "Monster Soup Commonly Called Thames Water, Being a Correct Representation of that Precious Stuff Doled Out to Us!!!" (http://www.scienceandsociety.co.uk/results. asp?image=10315337) Snow's achievements in improving sanitation to reduce cholera epidemics ultimately resulted in numerous innovations, including an effective sewage system, water filtration, and flush toilets (Keusch 2013).

A clean water supply addressed only one facet of food safety. In the late nineteenth century, some foodborne infections were also linked to increased bacterial growth in meat and dairy products when stored at room temperature. Refrigeration of food to less than 40 F slows the growth of pathogenic bacteria, which can grow rapidly at warmer temperatures without any evidence of spoilage to the consumer (http://www.fsis.usda. gov/shared/PDF/Refrigeration_and_Food_Safety.pdf). In 1882, William Soltau Davidson successfully pioneered the ability to transport frozen meat on trade ships from New Zealand to Great Britain (Palmer M). Davidson enlisted the *Bell-Coleman Mechanical Refrigeration Co.*, to equip a former passenger ship, the *Dunedin*, with steam-powered compression refrigeration machinery. This advancement was more cost-efficient than transporting live, and it allowed a large amount of meat to be transported for a prolonged period of time (Newsom 2006). This development eventually led to home refrigerators with freezers, thus allowing an increased longevity of purchased food products (Keusch 2013, Newsom 2006).

Another technique which prolongs the shelf life of various food products is pasteurization. This method was developed by Louis Pasteur, a French chemist who was commissioned by the French emperor Napoleon III, to research wine and beer impurities that were affecting storage and exportation negatively. In 1865, Pasteur discovered that heating wine resulted in a reduction of parasitic and fungal contamination (Debre 1994, English Translation 1998).

Currently, pasteurization is typically applied to dairy production. It involves heating food at a specified temperature and for a designated length of time to kill harmful bacteria, followed by prompt refrigeration to help prevent recontamination (2012 Department of Food, Nutrition, and Packaging Science, Clemson University). In early twentieth century Dr. Milton Rosenau, a US public health epidemiologist, was a strong proponent of milk pasteurization. In his book *The Milk Question*, he remarked that "raw milk is apt to be dangerous milk. Pasteurization is our only safeguard against certain of the dangers conveyed in milk ... The numerous outbreaks of typhoid fever, scarlet fever, diphtheria, sore throat, as well as the relation of milk to tuberculosis and other infections, is sufficiently real, sufficiently frequent, and sufficiently serious to arouse sanitarians and the public to a realization of the danger" (Rosenau 1912).

However, milk pasteurization was slow to adopt in the US in part because of the effect high temperatures had on the taste of milk. Dr. Rosenau discovered that "low temperature, slow pasteurization" eliminated bacteria but preserved the flavor. Public acceptance of pasteurized milk subsequently increased in the US (CDC 1999). The Grade A Pasteurized Milk Ordinance was enacted by the US Public Health Service in 1924 to help establish guidelines for milk sanitation, and it has been revised several times since then. In 1957, the ordinance was updated with the recommended parameters for destroying *Coxiella burnetti*, the agent causing Q fever in humans, and currently recognized as the most heat-resistant organism found in milk (http://www.milkfacts.info/Milk%20 Processing/Heat%20Treatments%20and%Pasteurization.htmm). Although pasteurization is commonly associated with dairy products, methods have also been developed for a variety of other food products. For example, steam pasteurization eliminates harmful bacteria including *E. coli* O157:H7, *Salmonella*, and *Listeria*, in almost half of freshly killed US beef. In addition, irradiation pasteurization employs gamma rays to reduce bacterial growth in meat, poultry, and produce (2012 Department of Food, Nutrition, and Packaging Science, Clemson University).

During the twentieth century, major policies regarding food safety in the United States were in part shaped by media popularization of outbreaks of foodborne illnesses as well as reports of unsanitary practices in food production factories. For example, Upton Sinclair, a journalist, investigated the unsanitary practices of the meatpacking industry in Chicago. He published his findings in *The Jungle* in 1906. His novel prompted a federal investigation of the meatpacking industry and threatened the country's trust in the safety of meat production (Barkan 1985, Kantor 1976). Later that year, Theodore Roosevelt passed The Pure Food and Drugs Act, requiring accurate labeling of unadulterated food products transported across states (U.S. Food and Drug Administration).

The spread of typhoid fever in early 1900s New York, and subsequent involuntary quarantine of Mary Mallon, serves as another example of how the management of foodborne illness outbreaks has been affected by prevailing public health concerns. Also known as "Typhoid Mary", Ms. Mallon was an asymptomatic carrier of *Salmonella typhi* and was implicated in 57 cases and 3 deaths from typhoid fever (American Journal of Public Health 1939). *Salmonella typhi* is transmitted by the fecal-oral route.

Prior to the development of chloramphenicol in 1948, typhoid fever had a 12% case-fatality rate (Chan and Reidpath 2003). Ms. Mallon was hired as a cook at several homes in New York. After several members in one household became infected with typhoid fever, George Soper from the New York Department of Health was enlisted to help find the source. He identified Ms. Mallon as a common connection between multiple affected households. She was subsequently escorted to the hospital where it was confirmed that she was shedding bacteria. Her gall bladder was suspected to be colonized by *Salmonella typhi*, allowing for chronic infection. Recommendations to prevent transmitting the disease were to retire from her job as a personal cook or to have her gall bladder removed. She refused both options, and officials had her isolated in quarantine until her death 26 years later (Brooks 1996).

Overview

Historically, food was produced and consumed locally, thereby limiting the opportunities for food contamination. But with industrialization and the need to feed growing populations, food production and processing has become increasingly centralized with a wider range of distribution. Following World War II, development of new technologies such as vacuum packing, better preservatives, deep freezing and microwaves allowed for mass production of shelf-stable foods with increased ease of preparation and consumption (Cutler et al. 2003). Subsequently, as food processing and eating habits changed, so have the predominant organisms causing foodborne illness. Prior to 1960, *Salmonella* spp., *Shigella* spp., *Clostridium botulinum* and *Staphylococcus aureus* were the major causes of gastrointestinal disease. *Clostridium perfringens* and *Bacillus cereus* emerged subsequently during the 1960s and were followed by norovirus, rotavirus and *Campylobacter* in the 1970s (Newell et al. 2010). In 1982, *E. coli* O157:H7 was first documented as a foodborne pathogen during an investigation of an outbreak of hemorrhagic colitis associated with hamburger consumption at a fast food restaurant chain (Riley et al. 1983). Around this time, *Cryptosporidia* and *Listeria* were also recognized as significant causes of foodborne illness (Newell et al. 2010). Though the first case of *Cyclospora* infection was reported in 1979, the 1996 outbreak of cyclosporosis associated with raspberries grown in Guatemala, heralded it as an emerging foodborne pathogen (Ortega and Sanchez 2010). This case focused the media spotlight on the growing reliance of U.S. on imported food and the potential for further outbreaks originating from abroad. New and surprising foodborne infections continue to emerge, with zoonoses and antibiotic-resistant bacteria at the forefront (Koluman and Dikici 2013, Newell et al. 2010).

Since 1996, the Foodborne Diseases Active Surveillance Network (FoodNet) has monitored the incidence of laboratory confirmed infections caused by nine commonly foodborne pathogens at 10 US sites. This collaboration between the Center for Disease Control, the Food Safety and Inspection Service of the US Department of Agriculture (USDA) and the Food and Drug Administration (FDA) has helped to quantify foodborne infections and guide policy on food safety. In 2006, the World Health Organization (WHO) formed a group of international experts to "Estimate the Global Burden of Foodborne Diseases;" final results are anticipated in 2015 (http://www.who.int/foodsafety/foodborne_disease/ferg/en/). The challenge of documenting and tracing foodborne infections has become more complicated with international trade and travel. Current estimates suggest that approximately 15% of the US food supply is imported from other countries (Taylor 2011). The Food Safety Modernization Act (FSMA) of 2011 represents another major safeguard against foodborne illnesses. The FSMA requires companies to develop measures for preventing foodborne illness outbreaks and provides the FDA authority for ensuring that these standards are upheld (Taylor 2011).

While significant strides have been made in the area of food safety, numerous challenges remain in the prevention and detection of foodborne illnesses. With the centralization of food production there is increased risk of contaminated food causing geographically widespread outbreaks. Similarly, imported foods and ingredients require reliance on appropriate food safety measures in the originating country, during transport and subsequent distribution. Emphasis on healthy eating corresponds to the growing foodborne illness outbreaks linked to fresh fruits and vegetables (Painter et al. 2013). New food vehicles, including processed and frozen foods and spices, are continually being identified (Behravesh et al. 2012) and previously controlled pathogens, such as *Clostridium botulinum*, have re-emerged (Juliao et al. 2013). Lifestyle and convenience results in a significant proportion of meals being eaten outside the home or as "take-out", and demand for food prepared by street vendors or food trucks may increase the risk of foodborne illness (Nyachuba 2010). Finally, the number of people susceptible to foodborne infections is growing as populations age and once fatal illnesses become treatable. Immunocompromised individuals, such as those affected by HIV, cancer, cirrhosis and transplants recipients are all at increased risk for infection, including those that are foodborne (Lund and O'Brien 2011). Gastric acid protects against certain enteric pathogens such as *Vibrio cholera* (Howden and Hunt 1987). Achlorhydria and hypochlorhydria, seen with malnutrition or acid-suppressing medications, may therefore predispose individuals to infectious diarrhea (Howden and Hunt 1987). Antacid use, particularly proton pump inhibitors, make individuals more susceptible to foodborne pathogens (Lund and O'Brien 2011). As these medications are available over-the-counter and widely used, consumers are unwittingly increasing their risk for foodborne infection.

Conclusion

Foodborne illnesses are a significant public health problem worldwide. Their heterogeneity, scope of disease, and changing epidemiology have necessitated the development of new surveillance methods and diagnostic tools. They affect a diverse and widespread population, with growing recognition as a serious threat to immunocompromised hosts. The impact of foodborne infections on our global society cannot be underestimated.

Keywords: foodborne illness, outbreak, public health, food safety

References

Anderson, R. 2011. Bugs Through the Ages: The Foodborne Illness Fight. Food Safety News.

Barkan, I.D. 1985. Industry invites regulation: the passage of the Pure Food and Drug Act of 1906. Am. J. Public Health. 75: 18–26.

Behravesh, C.B., I.T. Williams and R.V. Tauxe. 2012. Emerging Foodborne Pathogens and Problems: Expanding Prevention Efforts Before Slaughter or Harvest in Improving Food Safety through a One Health Approach: Workshop Summary. Washington, DC National Academies Press (US).

Berger, C.N., S.V. Sodha, R.K. Shaw, P.M. Griffin, D. Pink, P. Hand and G. Frankel. 2010. Fresh fruit and vegetables as vehicles for the transmission of human pathogens. Environ. Microbiol. 12: 2385–2397.

Brooks, J. 1996. The sad and tragic life of Typhoid Mary. CMAJ. 154: 915–916.

Centers for Disease Control and Prevention (CDC). 1999. Safer and healthier foods. MMWR Morb Mortal Wkly Rep. 48: 905–913.

Centers for Disease Control and Prevention (CDC). 2013. Surveillance for foodborne disease outbreaks—United States, 2009–2010. MMWR Morb Mortal Wkly Rep. 62: 41–47.

Chan, K.Y. and D.D. Reidpath. 2003. "Typhoid Mary" and "HIV Jane": responsibility, agency and disease prevention. Reprod Health Matters. 11: 40–50.

Cutler, D.M., E.L. Glaeser and J.M. Shapiro. 2003. Why have Americans become more obese? J. Econ. Perspect. 17: 93–118.

Debre, P. 1994. English Translation 1998. Louis Pasteur: Johns Hopkins University Press.

Department of Food, Nutrition, and Packaging Science, Clemson University. Describe Pasteurization. 2012. http://www.foodsafetysite.com/educators/competencies/general/foodprocessing/processing2.html (accessed 1/26/15).

Hempel, S. 2013. John Snow. Lancet. 381: 1269–1270.

Howden, C.W. and R.H. Hunt. 1987. Relationship between gastric secretion and infection. Gut. 28: 96–107.

http://www.ers.usda.gov/data-products/cost-estimates-of-foodborne-illnesses.aspx#. VDW27r4mUfy. USDA Economic Research Service. Cost Estimates of Foodborne Illness. (accessed 11/12/14).

http://www.fsis.usda.gov/shared/PDF/Refrigeration_and_Food_Safety.pdf (accessed 1/26/15).

http://www.milkfacts.info/Milk%20Processing/Heat%20Treatments%20and% Pasteurization.htmm (accessed 1/26/15).

(http://www.scienceandsociety.co.uk/results.asp?image=10315337) (accessed 2/3/15).

http://www.who.int/foodsafety/foodborne_disease/ferg/en/ (accessed 3.24.15).

Juliao, P.C., S. Maslanka, J. Dykes, L. Gaul, S. Bagdure, L. Granzow-Kibiger, E. Salehi, D. Zink, R.P. Neligan, C. Barton-Behravesh, C. Luquez, M. Biggerstaff, M. Lynch, C. Olson, I. Williams and E.J. Barzilay. 2013. National outbreak of type a foodborne botulism associated with a widely distributed commercially canned hot dog chili sauce. Clin. Infect. Dis. 56: 376–382.

Kantor, A.F. 1976. Upton Sinclair and the Pure Food and Drugs Act of 1906. I aimed at the public's heart and by accident I hit it in the stomach. Am. J. Public Health. 66: 1202–1205.

Keusch, G.T. 2013. Perspectives in foodborne illness. Infect. Dis. Clin. North Am. 27: 501–515.

Koluman, A. and A. Dikici. 2013. Antimicrobial resistance of emerging foodborne pathogens: status quo and global trends. Crit. Rev. Microbiol. 39: 57–69.

Kuchenmüller, T., S. Hird, C. Stein, P. Kramarz, A. Nanda and A.H. Havelaar. 2009. Estimating the global burden of foodborne diseases—a collaborative effort. Euro Surveill. 14.

Lund, B.M. and S.J. O'Brien. 2011. The occurrence and prevention of foodborne disease in vulnerable people. Foodborne Pathog. Dis. 8: 961–973.

Mallon Mary (Typhoid Mary). 1939. Am. J. Public Health Nations Health. 29: 66–68.

Markel, H. 2013. A piece of my mind. Happy birthday, Dr. Snow. JAMA. 309: 995–996.

McHugh, J. and P.A. Mackowiak. 2014. Death in the White House: President William Henry Harrison's Atypical Pneumonia. Clin. Infect. Dis. 59: 990–995.

Newell, D.G., M. Koopmans, L. Verhoef, E. Duizer, A. Aidara-Kane, H. Sprong, M. Opsteegh, M. Lanelaar, J. Threfall, F. Scheutz, J. van der Giessen and H. Kruse. 2010. Food-borne diseases—the challenges of 20 years ago still persist while new ones continue to emerge. Int. J. Food Microbiol. 139 Suppl. 1: S3–15.

Newsom, S.W. 2006. Pioneers in infection control: John Snow, Henry Whitehead, and Broad Street pump, and the beginnings of geographical epidemiology. J. Hosp. Infect. 64: 210–216.

Nyachuba, D.G. 2010. Foodborne illness: is it on the rise? Nutr. Rev. 68: 257–269.

Oldach, D.W., R.E. Richard, E.N. Borza and R.M. Benitez. 1998. A mysterious death. N. Engl. J. Med. 338: 1764–1769.

Ortega, Y.R. and R. Sanchez. 2010. Update on Cyclospora cayetanensis, a food-borne and waterborne parasite. Clin. Microbiol. Rev. 23: 218–234.

Painter, J.A., R.M. Hoekstra, T. Ayers, R.V. Tauxe, C.R. Braden, F.J. Angulo and P.M. Griffin. 2013. Attribution of foodborne illnesses, hospitalizations, and deaths to food commodities by using outbreak data, United States, 1998–2008. Emerg. Infect. Dis. 19: 407–415.

Palmer, M. 1882. William Soltau Davidson: A Pioneer of New Zealand Estate Management.

Riley, L.W., R.S. Remis, S.D. Helgerson, H.B. McGee, J.G. Wells, B.R. Davis, R.J. Hebert, E.S. Olcott, L.M. Johnson, N.T. Hargrett, P.A. Blake and M.L. Cohen. 1983. Hemorrhagic colitis associated with a rare *Escherichia coli* serotype. N. Engl. J. Med. 308: 681–685.

Rosenau, M.J. 1912. The Milk Question: Houghton Mifflin Company.

Scallan, E., P.M. Griffin, F.J. Angulo, R.V. Tauxe and R.M. Hoekstra. 2011a. Foodborne illness acquired in the United States—unspecified agents. Emerg. Infect. Dis. 17: 16–22.

Scallan, E., R.M. Hoekstra, F.J. Angulo, R.V. Tauxe, M.A. Widdowson, S.L. Roy, J.L. Jones and P.M. Griffin. 2011b. Foodborne illness acquired in the United States—major pathogens. Emerg. Infect. Dis. 17: 7–15.

Taylor, M.R. 2011. Will the Food Safety Modernization Act help prevent outbreaks of foodborne illness? N. Engl. J. Med. 365: e18.

US Food and Drug Administration. http://www.fda.gov/AboutFDA/WhatWeDo/History/Origin/ucm054819.htm (accessed 1/31/15).

CHAPTER 2

Recent Foodborne Infectious Diseases Outbreaks in the United States

Christian Rojas-Moreno[1],* and *William Salzer*[2]

Introduction

It is estimated that each year in the United States (U.S.), infectious agents cause 9.4 million foodborne illnesses: 5.5 million (59%) foodborne illnesses are caused by viruses, 3.6 million (39%) by bacteria and 0.2 million (2%) by parasites (Scallan et al. 2011). The pathogens that cause most illnesses are norovirus (58%), nontyphoidal *Salmonella* spp. (11%), *C. perfringens* (10%), and *Campylobacter* spp. (9%). Although only a small proportion of these illnesses occur in the setting of outbreaks, outbreak investigations provide valuable information that can help regulatory agencies develop strategies that enhance food safety and prevent further outbreaks related to specific pathogens and foods (CDC 2014).

A foodborne disease outbreak is defined as the occurrence of two or more cases of a similar illness resulting from ingestion of a common food. The Centers for Disease Control and Prevention (CDC) conducts foodborne disease outbreak surveillance in the United States through the Foodborne Disease Outbreak Surveillance System (FDOSS) (www.cdc.gov/foodsafety/fdoss/surveillance), a program for collection and periodic reporting of data

[1] Assistant Professor of Clinical Medicine, Division of Infectious Diseases, Department of Medicine, University of Missouri, USA.
[2] Professor of Medicine and Division Director, Division of Infectious Diseases, Department of Medicine, University of Missouri, USA.
* Corresponding author: rojasch@health.missouri.edu

on the occurrence and causes of foodborne disease outbreaks in the United States. State, local and territorial public health departments voluntarily submit reports of outbreaks investigated by their agencies using a web-based reporting platform, the National Outbreak Reporting System (NORS) (www.cdc.gov/nors). For foodborne outbreaks, NORS interfaces with FDOSS to collect information, such as:

- Date and location of the foodborne outbreak
- Number of people who became ill and their symptoms
- Food implicated in the outbreak
- Setting where the food was prepared and eaten
- Pathogen that caused the outbreak

The surveillance team analyzes the outbreak data and then makes it available using an online resource, the Food Outbreak Online Database (FOOD) (wwwn.cdc.gov/foodborneoutbreaks/). This tool provides access to national information and is intended to be used for limited descriptive summaries of outbreak data. It is available to anyone who may have questions about foodborne outbreaks in the United States. A listing and more detailed information about multistate outbreaks where the CDC was the lead public health agency is available at the Selected Multistate Foodborne Outbreak Investigation website (www.cdc.gov/foodsafety/outbreaks/multistate-outbreaks). This site also provides a link to the Morbidity and Mortality Weekly Report (MMWR) reports on foodborne illness and outbreaks (www.cdc.gov/foodsafety/outbreaks/multistate-outbreaks/reports.html).

Other helpful resources in the study of foodborne outbreaks in the United States are:

- FoodNet (Foodborne Diseases Active Surveillance Network) (www.cdc.gov/foodnet/) is America's report card for food safety over a period of time. This database is useful in showing disease trends. Foodborne diseases monitored through FoodNet include infections caused by *Campylobacter*, *E. coli* O157, *Listeria*, *Salmonella*, *Shigella*, *Vibrio*, and *Yersinia*, and parasites *Cryptosporidium* and *Cyclospora*.
- CaliciNet (National Norovirus Outbreak Network) (www.cdc.gov/norovirus/reporting/calicinet/) is a national norovirus outbreak surveillance network of federal, state, and local public health laboratories in the United States.
- PulseNet (National Laboratory Network) (www.cdc.gov/pulsenet/) is a network made up of 87 laboratories, at least one in each state. PulseNet connects foodborne illness cases together to detect and define outbreaks using DNA "fingerprinting" of the bacteria via pulsed-field

gel electrophoresis (PFGE). This tool is particularly useful in multistate outbreaks from a common source that has been widely distributed.

According to data from NORS, norovirus is the most common cause of confirmed, single-etiology outbreaks and illnesses, accounting for 41% of outbreaks in 2012. *Salmonella* was next, accounting for 25% of outbreaks. Of the 10,319 outbreak-related illnesses caused by a single confirmed etiologic agent, 7% resulted in hospitalization, with *Salmonella* causing the most outbreak-related hospitalizations. The most common food vehicles causing outbreaks were fish, vegetable row crops, and unpasteurized dairy products (CDC 2014).

Most Common Infectious Etiologies of Outbreaks in the United States

Bacteria

Salmonella (nontyphoidal)
Campylobacter jejuni
Vibrio parahaemolyticus
Listeria monocytogenes
Clostridium botulinum
Shiga toxin-producing *Escherichia coli*
Staphylococcus aureus
Bacillus cereus
Clostridium perfringens

Viruses

Norovirus
Sapovirus
Astrovirus

Parasites

Cyclospora
Giardia
Trichinella

This chapter will focus on the most recent and notable outbreaks caused by the most prevalent infectious etiologies and the food sources involved. Additionally, for each organism there will be a brief description, critical elements of their epidemiology and clinical features, as well as an updated list of outbreak vehicles.

Norovirus

Brief description: Norovirus, a member of the family *Caliciviridae*, is the leading cause of acute gastroenteritis in the United States across all age groups (Hall et al. 2013, Hall et al. 2014). There is great diversity among noroviruses, and human strains are classified on the basis of their genome sequences into 6 genogroups (GI–GVI), which are further divided into genotypes, with the prototype virus designated as GI.1 (i.e., genogroup I, genotype 1) (Glass et al. 2009, Leshem et al. 2013). Three genogroups (GI, GII and GIV) cause human disease and GII is currently the cause of most norovirus outbreaks in the United States. According to recently published data evaluating norovirus outbreaks in the United States between 2009 and 2012, GII caused 86% of norovirus outbreaks with GII.4 being the predominant genotype (Hall et al. 2014).

Epidemiology: According to the surveillance report from CDC, in 2012 norovirus caused 287 outbreaks (172 confirmed, 118 suspected), affecting 6,009 patients and causing 58 hospitalizations. However, the burden of norovirus disease in the United States is much larger. Hall et al. 2013, estimated that norovirus causes on average 570–800 deaths, 56,000–71,000 hospitalizations, 400,000 emergency room visits, 1.7–1.9 million outpatient visits, and 19–21 million total illnesses each year in the United States. This results in approximately $777 million in health-care costs.

Several characteristics of norovirus facilitate its spread (Glass et al. 2009): low infectious dose (18 to 1,000 viral particles) that allows transmission through droplets, fomites and person to person contact; viral shedding that precedes the onset of illness and may continue long after the illness (risk of secondary spread); virus tolerance to a wide range of temperatures with persistence on environmental surfaces, water and food items that are irrigated with contaminated water and eaten raw or undercooked; lack of complete cross-protection due to diversity of strains, leading to potential for repeated infections throughout life; and antigenic shift and recombination that results in new strains that are able to affect susceptible individuals.

Clinical features: The incubation period is 10 to 51 hours. Symptoms include vomiting, diarrhea, and sometimes fever, although norovirus infections can be asymptomatic (Division of Viral Diseases, Respiratory Diseases, and Prevention 2011). The illness normally lasts only 2 to 3 days but can last longer in nosocomial outbreaks, in immunocompromised adults and in children younger than 11 years of age (Glass et al. 2009). Rates of severe outcomes, such as hospitalization and death, are greatest in children aged < 5 years and older adults aged > 65 years (Hall et al. 2013).

Outbreak sources: Restaurants are the most common setting for foodborne norovirus outbreaks. Food handlers infected with norovirus are the commonest source of food contamination (Hall et al. 2014). Data from outbreak investigations revealed that 92% of implicated foods were contaminated during preparation, and 75% were foods eaten raw (Hall et al. 2014). The food categories most often attributed to outbreaks were vegetable row crops (e.g., lettuce and other leafy vegetables), fruits and mollusks, particularly oysters. Oyster-associated norovirus outbreaks often result from contamination at the source due to fecal contamination of growing waters.

Recent outbreak: In December 2009, over 200 individuals reported gastrointestinal symptoms after dining at a North Carolina restaurant (Alfano-Sobsey et al. 2012). The outbreak investigation found that the median incubation period was 25 hours and the median duration of illness was 24 hours. The most common symptoms were diarrhea (92%), vomiting (85%), nausea (31%) and stomach cramps (23%). Fever was present in approximately 20% of cases. Compared with controls, cases were 13 times more likely to have eaten any oysters and 12 times more likely to have eaten specifically steamed oysters, making this food the most likely source. A fifth of the households reported secondary cases (attack rate of 14% among 70 susceptible household contacts). The causative strain was identified as GII.12. The investigation concluded that the oysters were contaminated at the harvest site prior to arrival to the restaurant and that final cooked temperatures were inadequate to inactivate the virus. The lesson learned from this outbreak was that undercooked contaminated oysters pose a similar risk for norovirus illness as raw oysters.

Salmonella

Brief description: After norovirus, *Salmonella* is the second most common etiology of infectious foodborne outbreaks (CDC 2014). *Salmonella* is a gram-negative rod in the family Enterobacteriaceae that can cause a spectrum of illnesses in humans. The nomenclature of *Salmonella* is complex. The genus *Salmonella* contains two species, each of which contains multiple serotypes (Brenner et al. 2000). The two species are *S. enterica* and *S. bongori*. The serotypes can be names (such as Enteritidis, Typhimurium, Typhi, and Choleraesuis) or antigenic formulas. For named serotypes, to emphasize that they are not separate species, the serotype name is not italicized and the first letter is capitalized. *Salmonella* Typhi and *Salmonella* Paratyphi are the etiologic agents of enteric (typhoid) fever and have no known hosts other than humans. Most commonly, foodborne or waterborne transmission occurs as a result of fecal contamination by ill or asymptomatic chronic carriers. Unlike *S*. Typhi and *S*. Paratyphi, whose only reservoir is humans, nontyphoidal *Salmonella* can be acquired from multiple animal reservoirs that carry the organism and transmission can occur by consumption of animal products, contact with animals or their environment, and contaminated water. Nontyphoidal *Salmonella* can cause gastroenteritis, bacteremia and vascular infection (Mandal and Brennand 1988, Saphra and Winter 1957, Parsons et al. 1983).

Epidemiology: The CDC estimates that 1,027,561 cases of domestically acquired nontyphoidal salmonellosis occur annually in the U.S., when under-reporting and under-diagnosis are taken into account (Scallan et al. 2011). In 2012, 113 outbreaks of *Salmonella* were reported to NORS (106 confirmed, 7 suspected), causing 3,394 illnesses and 454 hospitalizations (the highest number of hospitalizations related to single etiology outbreaks). *Salmonella* also caused the highest number of multistate outbreaks. *Salmonella* Enteritidis was the most common (26%), followed by *Salmonella* Typhimurium (13%), *Salmonella* Newport (10%), *Salmonella* Javiana (7%) and *Salmonella* Heidelberg (6%). In contrast to nontyphoidal salmonellosis, typhoid fever in the United States is rare, with approximately 350 infections reported annually (Lynch et al. 2009). Researchers from the CDC identified 28 outbreaks of typhoid fever between 1999 and 2010 (Imanishi et al. 2014); most of them were linked to a confirmed or suspected carrier, but two were linked to imported frozen mamey fruit pulp and one was linked to Gulf Coast oysters.

Clinical features: In the case of nontyphoidal salmonellosis, the onset of illness is usually 6–72 hours after ingestion. Symptoms include nausea, vomiting, abdominal cramps, diarrhea, fever and headache and they usually last 4–7 days. In most cases, stools are loose, of moderate volume, and without blood. Enteric fever is a severe systemic illness characterized by fever and abdominal pain that is caused by *S*. Typhi and *S*. Parathyphi. The incubation period for *S*. Typhi is 10–14 days (range 5–60 days). Diarrhea occurs but constipation can affect a third of patients. Relative bradycardia (when the heart rate does not rise as expected for the increase in temperature) occurs in fewer than 50% of patients. Up to a third of patients can have the classic rose spots, a faint salmon-colored maculopapular rash on the trunk. The mortality rate of enteric fever without therapy is 10 to 15% but prompt treatment reduces that rate to less than 1%.

Outbreak sources: The most common implicated food categories in outbreaks of nontyphoidal salmonellosis in 2012 were fruits, fish and chicken (CDC 2014). Other food sources include cantaloupes, ground beef, poultry, cucumbers, mangos, peanut butter, sprouts, romaine lettuce, tomatoes, peppers, tuna, eggs, milk and dairy products, shrimp, spices, yeast, coconut, sauces, salad dressings made with unpasteurized eggs, cake mixes, cream-filled desserts and toppings that contain raw egg, dried gelatin, cocoa, and chocolate (FDA 2012).

Recent outbreaks: Between November 2008 and April 2009, 714 persons in 46 states were infected with *Salmonella* Typhimurium (Centers for Disease and Prevention 2009). The outbreak investigation started with an epidemiologic assessment of a growing cluster of isolates of *Salmonella* Typhimurium that shared the same pulsed-field electrophoresis (PFGE) pattern in PulseNet. The median age of patients was 16 years; 21% were aged < 5 years, and 15% were aged > 59 years. Fifty-two percent of the patients were male. Twenty-four percent of the patients were hospitalized, and eight patients aged ≥ 59 years might have died as consequence of the infection. Initial investigations found that peanut butter was among the most frequently reported food exposures in the 7 days before illness. By December 2008, the Minnesota Department of Health (MDH) learned from patient interviews that several affected patients lived or ate meals in one of three institutions that had a common food distributor and that the only food common to the three institutions was King Nut creamy peanut butter. The next month, six additional cases in six other institutions were identified and each of those institutions received King Nut peanut butter. Culture of the King Nut peanut butter sample collected from one of the institutions grew *Salmonella* Typhimurium, and the strain was later confirmed as the outbreak strain. King Nut peanut butter was produced by Peanut Corporation of America (PCA) at a single facility in Blakely, Georgia. PCA also produced peanut paste that was later associated with the outbreak when patient interviews indicated that many patients did not eat peanut butter in institutions, but had eaten various other peanut butter-containing products. Two cracker brands associated with the outbreak (Austin and Keebler), were both made at one plant which was known to receive peanut paste from PCA. This outbreak triggered the most extensive food recall in U.S. history and PCA was forced out of business. Findings from Food and Drug Administration (FDA) inspections of the PCA facility included: peanut-butter products contaminated with *Salmonella* but shipped anyway after "re-testing", lack of peanut paste line cleaning after isolation of *Salmonella*, failure to establish or record the effectiveness of the temperature in the roasting step, totes of raw peanuts stored directly next to totes of finished, roasted peanuts in the facility's production/packaging room, foot-long gaps in its roof, water stains on the ceiling and edges of skylights where rain water had been leaking into the facility (FDA 2009).

In 2010, the Southern Nevada Health District detected an outbreak of typhoid fever among persons who had not recently traveled abroad (Loharikar et al. 2012). The outbreak investigation revealed that consumption of frozen mamey fruit pulp was associated significantly with the illness. The majority of the patients reported consuming frozen mamey pulp milkshakes (*batidos*) in the 60 days before illness onset. Mamey is a tropical fruit grown primarily in Central and South America. Traceback investigations implicated *Goya* and *La Nuestra* brands of frozen mamey fruit pulp (from a single manufacturer in Guatemala) and they were recalled. The total count was 12 cases in 3 states. The median age of case patients was 18 years, 4 (33%) were male, and 11 (92%) were Hispanic. Nine patients (82%) were hospitalized; none of the patients died. Mamey pulp is vulnerable to contamination because its production involves hand manipulation at multiple points during processing. Inspections found deficient conditions at the manufacturing plant that could have allowed contamination by a convalescent or chronic asymptomatic carrier of *Salmonella* Typhi working at the plant. The alternative explanation was contamination through fecal contamination of the water supply to the plant.

Campylobacter

Brief description: *Campylobacter* spp. are gram-negative rods with curved to S-shaped morphology that cause gastroenteritis worldwide. They are commensal organisms routinely found in poultry, cattle, sheep and swine (Silva et al. 2011). The main species is *Campylobacter jejuni*.

Epidemiology: *Campylobacter* spp. are estimated to be the third leading cause of bacterial food illness in the United States. CDC estimates that each year there are approximately 800,000 cases of gastroenteritis due to *Campylobacter* spp., leading to 120,000 hospitalizations and 76 deaths (Scallan et al. 2011). In 2012, 37 outbreaks of *Campylobacter* spp. (30 confirmed, 7 suspected) were reported to NORS (CDC 2014); they caused 476 illnesses and 29 hospitalizations. The food category most often implicated in these outbreaks was unpasteurized dairy products, in contrast to sporadic illness by *Campylobacter* which is primarily associated with poultry products.

Clinical features: The disease caused by *Campylobacter jejuni* resembles salmonellosis clinically. The incubation period is 2 to 5 days. Symptoms include fever, diarrhea, abdominal cramps and vomiting. The diarrhea may be bloody. The disease is typically self-limited and lasts from 2 to 10 days (FDA 2012). A small percentage of patients, particularly those with immunosuppression, can develop complications such as bacteremia, meningitis, hepatitis, and pancreatitis. Other complications are related to post-infectious immune phenomena and include Guillain-Barré syndrome and reactive arthritis.

Outbreak sources: Common sources include improperly handled or undercooked poultry products, unpasteurized milk and cheeses and contaminated water. *C. jejuni* has been found in a variety of other foods, such as vegetables and seafood. *C. jejuni* is also found in surface water, such as the water found in ponds and streams (FDA 2012).

Recent outbreak: In October 2012, the Vermont Department of Health identified three cases of campylobacteriosis in Vermont residents (Centers for Disease and Prevention 2013); the isolates had indistinguishable pulsed-field gel electrophoresis (PFGE) patterns. The outbreak investigation, including a query of PulseNet, led to the identification of one additional case each from New Hampshire, New York, and Vermont that had been reported in the preceding 6 months. The total count was six patients; two of them required hospitalization. The investigations revealed that five patients were exposed through consumption of chicken livers and one patient worked at the establishment where the livers were processed. Livers collected from the processing establishment yielded the outbreak strain of *Campylobacter jejuni*. Although a food safety assessment found no major violations at the establishment, the company voluntarily ceased the sale of chicken livers. The lesson learned was that chicken livers are often contaminated internally with *Campylobacter* spp. and it is not safe to eat them raw or undercooked.

Shiga Toxin-producing Escherichia coli (STEC)

Brief description: *E. coli* is a gram-negative rod that belongs to the family Enterobacteriaceae. Most strains of *E. coli* are part of the normal gastrointestinal flora in humans. However, some *E. coli* are pathogenic and can cause intestinal and extra-intestinal disease. These pathogenic forms of *E. coli* are classified into pathotypes or groups that have a similar mode of pathogenesis and cause clinically similar forms of disease (Donnenberg and Whittam 2001). There are at least 8 recognized pathotypes of *E. coli*: enterotoxigenic (ETEC), enteropathogenic (EPEC), enteroinvasive (EIEC), enterohemorrhagic (EHEC), enteroaggregative (EAEC), diffusely adherent (DAEC), uropathogenic and meningitis-associated. The term STEC includes EHEC and non-EHEC strains that are capable of producing a Shiga-like toxin, which is the critical element in the pathogenesis of the disease caused by STEC (Nguyen and Sperandio 2012). STEC strains such as *E. coli* O157:H7 (STEC O157), have been implicated in outbreaks of foodborne diarrhea and its most serious complication, the hemolytic uremic syndrome (HUS), particularly in children. HUS is a microangiopathic hemolytic anemia characterized by disseminated capillary thrombosis and ischemic necrosis, with acute kidney injury as the consequence of renal involvement. One example of a non-EHEC Shiga toxin-producing strain is the enteroaggregative O104:H4 that caused the outbreak centered in Germany in 2011 (Frank et al. 2011). STEC O157 is the single serotype most frequently isolated and most often associated with HUS in the United States. However, non-O157 serotypes are actually more common than O157 as a group and can also cause foodborne outbreaks and serious illness (Hedican et al. 2009, Melton-Celsa et al. 2012).

Epidemiology: There are about 175,000 cases of STEC infections in the U.S. yearly (Scallan et al. 2011). Of those, approximately 112,000 are caused by STEC non-O157 and 63,000 are caused by STEC O157. A total of 29 foodborne outbreaks caused by STEC were reported in 2012 (all confirmed) and they caused 500 illness and 98 hospitalizations. STEC was second after *Salmonella* in the number of cases of hospitalization and the number of multistate outbreaks (CDC 2014). Of those 29 outbreaks, 24 were caused by STEC O157, 2 by STEC O145 and one each by STEC O45, O111 and O121. The serotypes O145, O45, O111 and O121 belong to a group of 6 serotypes (in addition to O26 and O103) that accounts for the majority of the non-O157 infections in the United States (Brooks et al. 2005).

Clinical features: The incubation period is usually 3–4 days but may range from 1 to 9 days. Infection can be asymptomatic or range from mild to severe with complications such as HUS. The acute presentation is typically hemorrhagic colitis, characterized by severe abdominal cramps and bloody diarrhea. HUS occurs in 5% to 10% of cases and is more common in young children and the elderly. Fever is typically low-grade or absent. Duration of symptoms is usually 2 to 9 days. In one study, cases of O157 infection were more likely to involve bloody diarrhea, hospitalization and HUS, compared with cases of non-O157 infection (Hedican et al. 2009).

Outbreak sources: Cattle are a natural reservoir of STEC (Hedican et al. 2009) (STEC is part of their normal flora), which is why raw or undercooked ground beef and beef products continue to be implicated in most infections. However, produce (contaminated with cow manure or water exposed to cow manure) has been increasingly implicated as a vehicle. In fact, the most common food category implicated in outbreaks in 2012 was vegetable row crops (CDC 2014). Outbreaks of STEC have been traced to sprouts, raw fruits such as apples or melons, unpasteurized milk, cookie dough, spinach, lettuce, yogurt, mayonnaise, fermented sausages and cheeses.

Recent outbreaks: On May 19, 2009, PulseNet identified a cluster of 17 cases of *E. coli* O157:H7 infections with indistinguishable PFGE patterns from 13 states (Neil et al. 2012). Early in the outbreak investigation, a clear hypothesis had not emerged, so a single interviewer conducted conversational open-ended interviews with 5 patients from Washington State to obtain detailed qualitative exposure histories and identify unusual exposures. They all reported consumption of raw commercial prepackaged cookie dough. Subsequent case-control studies found that cookie dough consumption was the only reported exposure significantly associated with illness. Although product and environmental investigations did not yield positive tests for the outbreak strain, the strong epidemiologic evidence implicating a specific brand of cookie dough as the vehicle for the infections led to a voluntary nationwide recall. The total count was 77 cases in 30 states. The median age of patients was 15 years; 66% were younger than 19 years of age. Twenty-one percent were male. Thirty-five of 64 (55%) patients with available information were hospitalized and 10 of 57 (18%) developed HUS; none of the patients died. A likely source of contamination was thought to be a contaminated ingredient such as flour, pasteurized eggs, chocolate chips, molasses, sugar, margarine, baking soda, and vanillin/vanilla extract. A lesson learned was that consumers often engage in risky behaviors such as eating unbaked products that are intended to be cooked before consumption.

In May 2011, public health authorities in Europe began investigating an outbreak of STEC O104:H4 infections that ultimately involved more than 4,000 persons in 16 countries, including the United States (Centers for Disease and Prevention 2013). On May 19, 2011, the Robert Koch Institute in Germany was notified of a cluster of three patients with HUS admitted to a single hospital in Hamburg. Enteroaggregative STEC O104:H4, a previously rare pathogen, was isolated from these patients. Over the next few weeks, case counts mounted rapidly in Germany, and new cases were quickly identified throughout Europe and elsewhere in persons who had recently traveled in Europe. Traceback investigations identified one lot of fenugreek seeds imported from Egypt as the source of the sprouts responsible from the outbreaks in Germany and France. One remarkable feature of this outbreak is that HUS affected mainly adults with a median age of 42 years; only 2% of the case patients with the HUS were younger than 5 years of age, as compared with 69% of case patients with HUS reported in Germany from 2001 through 2010 (Frank et al. 2011). This contrasts with HUS after STEC O157:H7 which predominantly affects children rather than adults (Boyce et al. 1995). In the United States, the CDC initiated active surveillance for cases associated with this outbreak on May 26. During the period May 26–June 16, six confirmed cases were identified in five states: Arizona, Massachusetts, Michigan, North Carolina, and Wisconsin. Cases were confirmed when STEC O104:H4 with a PFGE pattern matching the outbreak strain was isolated from a clinical specimen of a patient with pertinent travel history or contact with a person with a confirmed case. The median age of the patients was 52 years (range 38 to 72); four patients were male. All patients had diarrhea, including 4 with bloody diarrhea. Four patients (66%) developed HUS, requiring dialysis. One patient died. Five were primary cases that had traveled to Germany after April 1, 2011, and had the onset of illness during travel in Germany or within 3 weeks after returning from Germany; the additional secondary case was a close relative of a primary case. None of them recalled consumption of sprouts. After the health authorities in Germany declared the outbreak to be over, the final case count worldwide was 4,075 cases, of which 22% complicated by HUS, and 50 persons died.

Vibrio parahaemolyticus

Brief description: Vibrio parahaemolyticus, a member of *Vibrio* species from the Vibrionaceae family, is a halophilic gram-negative rod that is widely disseminated in estuarine, marine and coastal waters (Letchumanan et al. 2014). *V. parahaemolyticus* is transmitted by consumption of raw or undercooked seafood. It can also be transmitted by ingestion of any food contaminated by handling of raw seafood or by rinsing with contaminated water.

Epidemiology: The CDC estimates that there are approximately 34,000 foodborne illnesses caused by this organism each year (Scallan et al. 2011). Eleven outbreaks were reported to NORS in 2012 (8 confirmed, 3 suspected), with 66 reported illnesses and 5 hospitalizations. The most commonly implicated food category was mollusks (CDC 2014).

Clinical features: The incubation period is 4–90 hours and the duration of illness is 2–6 days (FDA 2012). Gastroenteritis can present with mild watery diarrhea or a frank, bloody diarrhea. Other symptoms include abdominal pain, nausea, vomiting, and fever. *V. parahaemolyticus* can also cause primary septicemia with mortality rate of 33% in one study (Hlady and Klontz 1996) liver disease was a common underlying condition in these cases.

Outbreak sources: The most common sources are raw or improperly cooked oysters. Other seafood products, including finfish, squid, octopus, lobster, shrimp, crab, and clams, have been linked to *V. parahaemolyticus* illnesses (FDA 2012). And as stated earlier, rinsing with contaminated water and handling raw seafood can convert any food into a potential source.

Recent outbreak: In August 2012, a *Vibrio parahaemolyticus* outbreak involving 6 persons occurred in Maryland, USA (Haendiges et al. 2014). The patients (members of 2 dining parties) had eaten in the same restaurant on the same day; raw and cooked seafood was served at the restaurant. The outbreak investigation did not identify the specific food responsible for the outbreak. The patients had not eaten oysters, lobster, or mussels, but they had eaten cooked clams, fish, crab, and shrimp. A traceback investigation was not conducted because the patients had not eaten oysters. The conclusion was that the outbreak was possibly caused by cross-contamination during food preparation. No other cases were reported from this restaurant or the surrounding area.

Listeria monocytogenes

Brief description: Listeria monocytogenes is a small gram-positive rod that rarely causes illness in the general population. But neonates, pregnant women, the elderly and immunosuppressed patients are at risk of life-threatening disease by this organism, including meningoencephalitis and bacteremia (Lorber 1997). It has been known for a long time that many patients experience diarrhea before the development of bacteremia or meningoencephalitis due to *Listeria monocytogenes*, but it was only in the late 90s that convincing evidence was obtained that this organism can cause acute, self-limited, febrile gastroenteritis in healthy persons (Dalton et al. 1997, Ooi and Lorber 2005). *Listeria monocytogenes* can grow at refrigerated temperatures, which makes this organism a problem for the food industry.

Epidemiology: The CDC estimates that there are 1,591 cases of foodborne listeriosis annually (Scallan et al. 2011), associated with 255 deaths. In 2012, 5 outbreaks were reported to NORS (4 confirmed, 1 suspected). They caused 42 illnesses and 38 hospitalizations (second to botulism in percentage of hospitalization) (CDC 2014). Among the 23 deaths due to foodborne outbreaks reported in 2012, *Listeria monocytogenes* was the most common etiology (6 deaths). Listeriosis is regarded as the most lethal cause of foodborne outbreaks.

Clinical features: Foodborne listeriosis can present in two ways: non-invasive gastroenteritis (self-limited in healthy individuals) and the invasive form that can lead to bacteremia, meningitis and death. The latter occurs predominantly in older adults, pregnant women, newborns and immunosuppressed persons. The non-invasive form has an incubation period of few hours to 2–3 days and the invasive form has a longer incubation period that varies from 3 days to 3 months (FDA 2012). The symptoms most frequently reported in gastroenteritis outbreaks are fever, arthralgia, myalgia, diarrhea and headaches (Ooi and Lorber 2005). Diarrhea is typically nonbloody and watery. Interestingly, sleepiness was reported in 63% of patients in a massive outbreak involving 1,566 students and staff of primary schools in Italy (Aureli et al. 2000).

Outbreak sources: Many foods have been associated with *L. monocytogenes* (FDA 2012): raw or inadequately pasteurized milk, chocolate milk, ice cream, cheeses (especially soft cheeses), raw vegetables, raw poultry and meats, fermented raw-meat sausages, hot dogs and deli meats, and raw and smoked fish and other seafood. More recently, cantaloupes and caramel apples have been associated with outbreaks.

Recent outbreaks: On September 2, 2011 the Colorado Department of Public Health and Environment (CDPHE) notified the CDC of a listeriosis outbreak (McCollum et al. 2013). Seven cases were reported in the last 3 days of August 2011, when two cases are typically reported each August in Colorado. PulseNet identified additional related isolates from patients residing in other states. At the end of the outbreak investigation, 147 outbreak-related cases were identified in 28 states. 99% of the patients were hospitalized and 22% died. The majority of patients were 60 years of age or older (86%). The median age was 77 years and the median age of those who died was 81 years. Most cases occurred in five adjoining states: Colorado (40), Texas (18), New Mexico (15), Oklahoma (12), and Kansas (11). Investigations revealed that patients with outbreak-related illness were significantly more likely to have eaten cantaloupe than were patients 60 years of age or older with sporadic listeriosis. Eight percent of the patients who recalled the cantaloupe brand, reported consuming regionally branded cantaloupe grown in Colorado. On September 10, 2011, the CDPHE and FDA conducted a joint inspection of the only farm in Colorado still engaged in cantaloupe processing that time of the year. Environmental swabs yielded *L. monocytogenes* isolates with PFGE patterns that were indistinguishable from three outbreak-related isolates, confirming the farm as the source of contaminated cantaloupes. It was concluded that the recent introduction of equipment to accommodate changes in cantaloupe washing and drying processes probably created environmental conditions that promoted cantaloupe contamination (a recirculating, chlorinated, chilled-water wash to clean and precool cantaloupes was changed in 2011 to a nonrecirculating wash method using municipal water, without supplemental chlorine, and a series of brash and felt rollers to mechanically clean and dry cantaloupes). The *Listeria* Initiative turned out to be a very useful tool in this outbreak. The *Listeria* Initiative is an enhanced surveillance system that collects reports of laboratory-confirmed cases of human listeriosis in the United States. The main objective of the *Listeria* Initiative is to help in the investigation of listeriosis clusters and outbreaks by decreasing the time from outbreak detection to public health intervention. Demographic, clinical, laboratory, and epidemiologic data are collected using a standardized, extended questionnaire. Interviews are conducted as cases are reported, rather than after clusters are identified, to minimize the effect of recall bias on food consumption history. Additionally, clinical, food, and environmental isolates of *L. monocytogenes* are subtyped using PFGE and the results are submitted to PulseNet. In this particular outbreak, the *Listeria* Initiative facilitated rapid case-case comparisons, which quickly confirmed the association between the outbreak-related illnesses and cantaloupe consumption, prompting the initial farm visit and same-day halt of cantaloupe production 12 days after recognition of the outbreak in Colorado.

More recently, the CDC collaborated with public health officials in several states and with the FDA to investigate an outbreak of

Listeria monocytogenes infections with commercially produced, prepackaged caramel apples as the likely source (CDC). Public health investigators used the PulseNet system to identify illnesses that were part of this outbreak. DNA fingerprinting was performed on *Listeria* isolated from ill people using PFGE and whole-genome sequencing (WGS). WGS gives a more detailed DNA fingerprint than PFGE. The 35 ill people included in this outbreak investigation were reported from 12 states: Arizona (5), California (3), Colorado (1), Minnesota (4), Missouri (5), Nevada (1), New Mexico (6), and Wisconsin (3). Eleven illnesses were associated with a pregnancy. One fetal loss was reported. Among people whose illnesses were not associated with a pregnancy, the median age of 62 years, and 33% were female. Three invasive illnesses (meningitis) occurred among otherwise healthy children aged 5–15 years. Thirty-four people were hospitalized, and listeriosis contributed to at least three of the seven deaths reported.

Clostridium perfringens

Brief description: Clostridium perfringens is an anaerobic, spore-forming, gram-positive rod found in many environmental sources as well as in the intestines of humans and animals. *C. perfringens* type A causes virtually all cases of clostridial food poisoning in the United States. Contaminated food products that are not properly cooked or stored allow for the proliferation of large numbers of vegetative *C. perfringens* cells. Once consumed, the vegetative cells sporulate within the small intestine, releasing an enterotoxin that causes the characteristic symptoms (Onderdonk and Garrett 2010). *C. perfringens* type C, the cause of pigbel (enteritis necroticans), produces a necrotizing small bowel inflammation that was seen in Papua New Guinea in the 1960s and 1970s prior to vaccination programs (Poka and Duke 2003).

Epidemiology: It is estimated that *C. perfringens* causes 965,958 foodborne illnesses each year in the U.S., second only to *Salmonella* when considering bacterial causes of foodborne illness (Scallan et al. 2011). There were 25 reported outbreaks in 2012 (18 confirmed), causing 1,062 illnesses and 4 hospitalizations (CDC 2014).

Clinical features: The incubation period for *C. perfringens* is generally between 7 and 15 hours with a range of 6 to 24 hours. In other words, this is the time needed for the vegetative cells to sporulate and release the toxin *in vivo*. Symptoms include watery diarrhea, abdominal cramping, vomiting, and fever. Cases resolve spontaneously within 24 to 48 hours (Onderdonk and Garrett 2010).

Outbreak sources: The food category most commonly implicated in outbreaks in 2012 was pork (CDC 2014). Other common sources are meats (especially beef and poultry), meat-containing products (e.g., gravies and stews), and vegetables, including spices and herbs (FDA 2012). In most instances, the actual cause of the intoxication is temperature abuse of cooked foods.

Recent outbreak: On May 7, 2010, 42 residents (out of 136, attack rate 31%) at a Louisiana state psychiatric hospital experienced vomiting, abdominal cramps, and diarrhea (Centers for Disease and Prevention 2012). Nine out of 13 staff members also experienced those symptoms (attack rate 69%). Illness onset ranged from 9 pm on May 6 through 3 pm on May 7. Because of the tight clustering of symptom onset, food served at the evening meal on May 6 was considered to be the most likely source of the outbreak. The mean incubation period from eating the suspect meal to symptom onset was 13 hours (range 5–21 hours). The most common symptoms were diarrhea (94%), abdominal cramps (51%) nausea (39%), and vomiting (27%). Interviews of the kitchen staff revealed that the chicken served at the suspect meal was delivered frozen to the kitchen on May 4, and was cooked on May 5, the day before serving. Contrary to hospital guidelines, the chicken was placed in 6-inch deep pans after cooking and covered with aluminum foil which slowed cooling, and the first temperature check was not done until 16 hours later. The state public health laboratory detected *C. perfringens* enterotoxin by reversed passive latex agglutination (RPLA) in 20 of 23 stool specimens from ill residents. The CDC also detected the *C. perfringens* enterotoxin gene in all four of the samples of chicken served at the suspect meal. Within 24 hours, three patients had died. The three fatalities occurred among patients aged 41–61 years who were receiving medications that decreased intestinal motility. For two of three decedents, the cause of death found on postmortem examination was necrotizing colitis. The laboratory results, clinical course, and epidemiologic findings indicated that this outbreak was caused by *C. perfringens* type A. They also suggested that psychiatric inpatients, especially those with decreased intestinal motility due to medications, are vulnerable to severe outcomes from *C. perfringens* type A intoxication.

Clostridium botulinum

Brief description: Clostridium botulinum is an anaerobic, spore-forming, gram-positive rod. Its potent neurotoxin causes botulism, a symmetric, descending and potentially lethal flaccid paralysis of motor and autonomic nerves. Botulinum toxins are the most potent toxins known; estimated lethal doses for purified crystalline botulinum toxin type A for a 70-kg man are 0.09–0.15 µg when introduced intravenously, 0.80–0.90 µg when introduced by inhalation and 70 µg when introduced orally (Sobel 2005). The toxin is usually quantified in terms of biological activity expressed as mouse intraperitoneal lethal dose ($MIPLD_{50}$). Botulinum toxins are designated types A through G based on antigenic differences. Human botulism in the U.S. is primarily caused by the strains of *C. botulinum* that produce toxin types A, B, and E (Shapiro et al. 1998). Type A has been associated with more severe disease and a higher fatality rate than type B or type E toxins. Foodborne botulism is caused by ingesting preformed *Clostridium botulinum* neurotoxin. The organism can cause food poisoning because the heat-resistant spores survive food preservation methods that kill nonsporulating organisms; they subsequently produce the neurotoxin under anaerobic, low-acid and low-solute conditions and the preformed toxin causes botulism after ingestion of the contaminated food. The canning and fermentation of foods, for example, create anaerobic conditions that allow *C. botulinum* to germinate (Sobel et al. 2004).

Epidemiology: From 1990 to 2000, 160 foodborne botulism events affected 263 people in the United States, an annual incidence of 0.1 per million (Sobel et al. 2004). The median number of cases per year was 23 (range 17–43). The most common toxin type was A. Home-canned foods were the leading cause. In 2012, 6 outbreaks of *Clostridium botulinum* were reported to NORS (5 confirmed, 1 suspected). There were 21 cases and all of them were hospitalized (CDC 2014).

Clinical features: The clinical syndrome of foodborne botulism is dominated by neurologic symptoms and signs resulting from a toxin-induced blockage of acetylcholine release (Shapiro et al. 1998). The incubation period is 18–36 hours (range 4 hours to 8 days). The initial symptoms may be gastrointestinal and can include nausea, vomiting, abdominal cramps, or diarrhea; after the onset of neurologic symptoms, constipation is more typical. Dry mouth, blurred vision, and diplopia are usually the earliest neurologic symptoms. These initial symptoms may be followed by dysphonia, dysarthria, dysphagia, and peripheral muscle weakness. Symmetric descending paralysis is characteristic; paralysis begins with the cranial nerves, and then affects the upper extremities, the respiratory muscles, and, finally, the lower extremities in a proximal-to-distal pattern. Respiratory muscle paralysis can lead to respiratory failure and death.

Outbreak sources: Foods associated with botulism include canned corn, peppers, green beans, beets, asparagus, mushrooms, spinach, tuna fish, chicken and chicken livers, luncheon meats, ham, sausage, lobster, smoked and salted fish (FDA 2012).

Recent outbreak: During October 2–4, 2011, eight maximum security inmates at the Utah State Prison in Salt Lake County were diagnosed with foodborne botulism (Centers for Disease and Prevention 2012). On October 2, 2011, a patient was hospitalized with a 3-day history of dysphagia, double vision, progressive weakness, and vomiting. He reported that his symptoms began within 12 hours of drinking pruno (an illicit alcoholic brew made of fruit, sugar, and water among other ingredients). Inmates who had consumed pruno or had symptoms of botulism were urged to seek medical attention. By October 4, 12 additional inmates sought medical attention for clinical complaints or history of recent pruno consumption. Of the 13 inmates who reported drinking pruno, eight met the case definition by having signs or symptoms compatible with botulism. These eight inmates were admitted to the neuro-critical care unit of hospital A and treated with heptavalent botulinum antitoxin (HBAT). Specimens from five of the eight confirmed patients tested positive for *C. botulinum* type A or its toxin. The patients were aged 24–35 years and lived in close proximity within the same maximum security prison unit. The median time to onset of symptoms was 37 hours after consumption of pruno brew A (range: < 12–80 hours). Three patients were placed on mechanical ventilation within 24 hours of admission. The total time spent in the hospital ranged from 2 to 58 days. After hospital discharge, all eight patients were evaluated in the prison infirmary where they received care for 1–76 days. Most of the inmates continued to have various clinical complaints 11 months after the outbreak, including weakness and loss of muscle mass, dysphagia and reflux. All the eight hospitalized patients drank pruno from the same batch (brew A) on September 30, 2011; in addition, two of the eight drank pruno from a second batch (brew B) on October 2, 2011. Of the five inmates who did not develop botulism, one reported tasting a small amount of brew A, which he spat out and four reported consuming only brew B. Although most ingredients used in the two brews were the same, a baked potato was included in brew A but not in brew B. A moist sock used to filter brew A was submitted for testing and a small amount of pruno squeezed out of the sock yielded *C. botulinum* type A. Pruno brew A was made with oranges, grapefruit, canned fruit, water, powdered drink mix (a source of sugar), and a baked potato. The inmate who prepared brew A reported the potato was removed from a meal tray, stored at ambient temperature for an undetermined number of weeks in either a sealed plastic bag or jar, peeled using his fingernails, and added to a plastic bag containing other ingredients a few days before brew A consumption. The ingredients were fermented in this bag for several days before being distributed to other inmates. Toxin was likely produced when the potato was added to the bag containing low acidity pruno ingredients under warm, anaerobic conditions during pruno fermentation.

Staphylococcus aureus

Brief description: Staphylococcus aureus is gram-positive coccus that can cause a wide range of infections in humans. Staphylococcal food poisoning, one of the most common foodborne illnesses in the United States, is caused by ingestion of preformed staphylococcal enterotoxins. These toxins are heat-stable so they are not destroyed by cooking.

Epidemiology: The CDC estimates that, in the United States, staphylococcal food poisoning causes approximately 241,188 illnesses, 1,064 hospitalizations, and 6 deaths each year (Scallan et al. 2011). In 2012, five outbreaks were reported to NORS (2 confirmed, 3 suspected) and they caused 149 illnesses and 4 hospitalizations (CDC 2014).

Clinical features: The incubation period is usually short (1–7 hours) and it also depends on individual susceptibility to the toxin, amount of toxin ingested and general health. Common symptoms include nausea, abdominal pain, vomiting and diarrhea. There is no fever. It is usually self-limited and lasts only a few hours to one day.

Outbreak sources: Foods frequently implicated in staphylococcal food poisoning include meat and meat products; poultry and egg products; salads, such as tuna, chicken, potato, and macaroni; bakery products, such as cream-filled pastries, cream pies, and chocolate éclairs; sandwich fillings milk and dairy products. Humans can be carriers of the bacterium and foods that require considerable handling during preparation and are kept slightly above proper refrigeration temperatures for an extended period predispose to bacterial proliferation and toxin production leading to staphylococcal food poisoning (FDA 2012).

Recent outbreak: On July 30, 2012, the emergency department at a military hospital was visited by 13 persons seeking care for gastrointestinal illness with onset 2–3 hours after a work lunch party (Centers for Disease and Prevention 2013). Twenty-two out of 40 attendees met the case definition. Among the 22 patients, 19 (86%) had nausea, 15 (69%) vomiting, 17 (77%) diarrhea, 17 (77%) abdominal pain, and 13 (59%) headache. Mean self-reported period to illness onset was 2.1 hours from the time of consumption, and mean duration of illness was 10.7 hours (range: 1–32 hours). Illness was associated with eating perlo (a chicken, sausage, and rice dish). The stool specimens from three patients and samples from the four main dishes (perlo, chicken wings, pulled pork, and green beans with potatoes) were sent to CDC for laboratory testing which detected staphylococcal enterotoxin A in the perlo dish. Testing for *B. cereus* and *C. perfringens* was negative. The initial source of contamination of the perlo is unknown but might have occurred while the preparer was handling the chicken after it was initially cooked. The precooked perlo was stored overnight in an unrefrigerated environment and this was the probable cause of organism proliferation and enterotoxin production. Rewarming of the perlo for approximately 1 hour the following day did not destroy the heat-stable toxin.

Bacillus cereus

Brief description: Bacillus cereus is an aerobic, spore-forming, gram-positive rod that is ubiquitous in the environment because the spores are very hardy and can survive for many years. The toxins of *B. cereus* can cause food poisoning of two types: the diarrheal type, associated with meat dishes and sauces, and the emetic type, associated with fried rice. Classically, the diarrheal type mimics *Clostridium perfringens* intoxication and the vomiting type mimics *Staphylococcus aureus* intoxication (FDA 2012). The pathogenesis in the diarrheal type involves toxin production *in vivo*, in contrast to the emetic type where food poisoning occurs after ingestion of a preformed heat-stable toxin.

Epidemiology: The estimated number of episodes of *B. cereus* illness annually is 63,400 (Scallan et al. 2011). Only two outbreaks were reported to NORS in 2012 (one confirmed, one suspected), causing 24 illnesses (CDC 2014). None of the reported cases required hospitalization.

Clinical features: The incubation period is 6–15 hours for the diarrheal type (range 1–24 hours) and 0.5–6 hours for the vomiting type. The diarrheal type is manifested by profuse diarrhea with abdominal pain and cramps; fever and vomiting are uncommon. The emetic type is manifested by nausea, vomiting, abdominal cramps, and occasionally diarrhea. In both types, symptoms usually resolve after 24 hours.

Outbreak sources: The diarrheal-type outbreaks have been associated with meats, milk, vegetables, and fish. The emetic-type outbreaks have generally been associated with rice products; however, other starchy foods, such as potato, pasta, and cheese products, have also been implicated (FDA 2012). Food mixtures, such as sauces, puddings, soups, casseroles, pastries and salads, have been linked to foodborne outbreaks.

Example of outbreak: Although not recent, the following is a good example of *B. cereus* outbreak from fried rice. On July 21, 1993, the Lord Fairfax (Virginia) Health District received reports of acute gastrointestinal illness that occurred among children and staff at two jointly owned child day care centers following a catered lunch (Centers for Disease and Prevention 1994). The catered lunch was served to 82 children and nine staff. A case was defined as vomiting by a person who was present at one of the two day care centers on July 21. The case control study among

80 interviewed people revealed that 14 out of 67 persons who had the lunch became ill, compared with none of 13 who did not. In addition to vomiting, the symptoms included nausea (71%), abdominal cramps or pain (36%), and diarrhea (14%). The median incubation period was 2 hours (range 1.5–3.5 hours). Symptoms resolved a median of 4 hours after onset (range 1.5–22 hours). Chicken fried rice prepared at a local restaurant was the only food significantly associated with illness; *B. cereus* was isolated from leftover chicken fried rice and from vomited material from one ill child. The rice had been cooked on the night of July 20 and cooled at room temperature before refrigeration. On the morning of the lunch, the rice was pan-fried in oil with pieces of cooked chicken, delivered to the day care centers at approximately 10:30 am, held without refrigeration, and served at noon without reheating. Vegetative forms of the organism probably multiplied and released the toxin at the restaurant and the day care centers while the rice was held at room temperature.

Other Notable Organisms that Caused Outbreaks Recently

Cyclospora cayetanensis

Brief description: Cyclospora cayetanensis, a protozoan parasite, is a cause of diarrhea worldwide, particularly in tropical regions. Persons at risk are travelers to endemic areas and those with acquired immunodeficiency syndrome (AIDS). In the United States, it has been associated with foodborne outbreaks.

Epidemiology: Cyclospora cayetanensis is endemic in Haiti, Peru, Nepal, Guatemala and other tropical and subtropical countries. In the United States, only few cases are reported in individuals that don't have a travel history or that are not related to an outbreak. It is estimated that approximately 11,000 cases of foodborne cyclosporiasis occur each year in the United States (Scallan et al. 2011). Several foodborne outbreaks, including multistate outbreaks, have been reported in the United States, with Guatemalan raspberries being one of the most commonly implicated vehicles (Ortega and Sanchez 2010). One biological characteristic of this organism that makes person to person transmission unlikely is that when excreted, the oocysts require 7–15 days outside the host to sporulate and become infectious.

Clinical features: The incubation period is approximately one week. Symptoms of cyclosporiasis include anorexia, nausea, flatulence, fatigue, abdominal cramping, diarrhea, low-grade fever and weight loss (Ortega and Sanchez 2010). The diarrhea is usually watery and can be protracted. AIDS patients may have more severe and more prolonged symptoms compared to non-AIDS individuals. Infection can be asymptomatic in endemic areas.

Outbreak sources: Vehicles for U.S. outbreaks were mainly imported fresh produce: raspberries, mesclun, snow peas, basil (Herwaldt 2000, Ortega and Sanchez 2010).

Recent outbreak: During June–August 2013, CDC, state and local public health officials, and the Food and Drug Administration (FDA) investigated an unusually large number of reports of cyclosporiasis (Centers for Disease and Prevention 2013). By September 20, CDC had been notified of 643 cases from 25 states, primarily Texas (278 cases), Iowa (153), and Nebraska (86). Results from outbreak investigations indicated that there was more than one outbreak of cyclosporiasis during June–August 2013 in the United States. Public health officials in Iowa and Nebraska performed investigations within their states and concluded that restaurant-associated cases of cyclosporiasis in their states were linked to a salad mix produced by Taylor Farms de Mexico. Epidemiologic and traceback investigations conducted in Texas by state and local public health officials, the FDA, and CDC indicated that some illnesses among Texas residents were linked to fresh cilantro from Puebla, Mexico.

Hepatitis A

Brief description: The hepatitis A virus is a non-enveloped single-stranded RNA virus that belongs to the picornavirus family. It causes acute inflammation of the liver. The virus is relatively resistant to heat and for complete inactivation food needs to be heated to > 85°C for at least one minute. Three genotypes infect humans (I, II, and III).

Epidemiology: Foodborne hepatitis A outbreaks are infrequent in the United States and it is estimated that every year there are around 1,500 cases of hepatitis A from consumption of contaminated food (Scallan et al. 2011). The rates of hepatitis A have been declining since the 1996 introduction of the hepatitis A childhood vaccination program.

Clinical features: Hepatitis A is acute, self-limited and doesn't become chronic. The incubation period is 15–50 days with an average of 28 days. Clinical presentation is indistinguishable from other causes of hepatitis: fever, dark urine, jaundice, malaise, nausea, vomiting, abdominal pain, and arthralgia. The illness typically lasts 2–3 weeks but individuals with chronic liver disease can have a fulminant course with acute liver failure, a complication that is rare in otherwise healthy persons. The infection is more commonly asymptomatic in children than in adults.

Outbreak sources: Contamination of food by infected workers in processing plants and restaurants is the common source of outbreaks. Specific vehicles include: cold cuts and sandwiches, fruits and fruit juices, milk and milk products, vegetables, salads, shellfish, and iced drinks (FDA 2012).

Recent outbreak: In May, 2013, an outbreak of symptomatic hepatitis A virus infections occurred in the USA (Collier et al. 2014). 165 people from ten states met the case definition. Dates of illness onset ranged from March 31, 2013 to August 12, 2013. 69 (42%) of 165 patients were hospitalized, two (1%) developed fulminant hepatitis, and one patient (a secondary case) needed a liver transplant; none of the patients died. Most patients were aged 40–64 years and most were female. A frozen mix of cherries, strawberries, raspberries, blueberries, and pomegranate from a retailer was associated with the outbreak and further investigation identified frozen pomegranate arils imported from Turkey as the source. Of 157 patients, 152 (97%) reported consumption or purchase of the product. Of the 165 patients, 120 had clinical specimens available for testing, including 119 serum samples and one stool specimen. 117 (98%) tested positive for hepatitis A genotype IB. State and local health departments gave post exposure prophylaxis to thousands of people. The involved retailer paid for vaccination for more than 10,000 people at more than 200 stores. In this case, public health action was not delayed to wait for results of a case-control study because hepatitis A has a long incubation period making patient recall of dietary exposures questionable, the shelf life of the implicated product was 2 years, and there was a continuing health risk. Although cherries and strawberries were also common among lots of the frozen mix, both the cherries and strawberries were used in many other products and no outbreak-associated cases were recorded among people who had consumed those products.

Conclusions

A foodborne disease outbreak is the occurrence of two or more cases of a similar illness resulting from ingestion of a common food. Numerous infectious etiologies, including bacteria, viruses and parasites cause outbreaks each year and are responsible for significant morbidity to the United States population.

This chapter highlights the key features of the most common causes of foodborne infectious outbreaks in the United States. The examples of recent outbreaks provide more detailed information about their epidemiology, the importance of outbreak investigations, and the role that CDC, FDA and state and local health departments play to ensure food safety. We can learn from these outbreaks: for example, a Shiga toxin-producing *E. coli* that is not O157:H7 (enteroaggregative *E. coli* O104:H4) causing a large outbreak that affected mainly adults and often complicated with HUS or a recent outbreak of *Listeria monocytogenes* where the investigators used newer molecular techniques such as whole-genome sequencing for epidemiologic purposes. It is important to keep in mind that there is a wide variety of vehicles for outbreaks. In this chapter, we describe the common sources of outbreaks in the United States, but we will always need to be vigilant for new vehicles of foodborne illnesses, such as the cookie dough that for the first time was associated to *E. coli* O157:H7 in the 2009 outbreak or the caramel apples that caused a multistate outbreak of listeriosis in 2014.

The CDC website has robust information about past outbreaks and updated information on recent and ongoing outbreaks, as well as information about product recalls. All that information gathered from recent and past outbreaks is used in the food industry to ensure quality and safety along the multiple steps that are involved in the process, to prevent future foodborne outbreaks.

Keywords: foodborne outbreaks, infectious diseases, outbreak investigation, outbreak vehicles, PulseNet

References

Alfano-Sobsey, E., D. Sweat, A. Hall, F. Breedlove, R. Rodriguez, S. Greene, A. Pierce, M. Sobsey, M. Davies and S.L. Ledford. 2012. Norovirus outbreak associated with undercooked oysters and secondary household transmission. Epidemiol. Infect. 140(2): 276–82.

Aureli, P., G.C. Fiorucci, D. Caroli, G. Marchiaro, O. Novara, L. Leone and S. Salmaso. 2000. An outbreak of febrile gastroenteritis associated with corn contaminated by *Listeria monocytogenes*. N. Engl. J. Med. 342(17): 1236–41.

Boyce, T.G., D.L. Swerdlow and P.M. Griffin. 1995. *Escherichia coli* O157:H7 and the hemolytic-uremic syndrome. N. Engl. J. Med. 333(6): 364–8.

Brenner, F.W., R.G. Villar, F.J. Angulo, R. Tauxe and B. Swaminathan. 2000. Salmonella nomenclature. J. Clin. Microbiol. 38(7): 2465–7.

Brooks, J.T., E.G. Sowers, J.G. Wells, K.D. Greene, P.M. Griffin, R.M. Hoekstra and N.A. Strockbine. 2005. Non-O157 Shiga toxin-producing *Escherichia coli* infections in the United States, 1983–2002. J. Infect. Dis. 192(8): 1422–9.

Centers for Disease, Control and Prevention. 1994. *Bacillus cereus* food poisoning associated with fried rice at two child day care centers—Virginia, 1993. MMWR Morb Mortal Wkly Rep. 43(10): 177–8.

Centers for Disease Control and Prevention. 2009. Multistate outbreak of Salmonella infections associated with peanut butter and peanut butter-containing products—United States, 2008–2009. MMWR Morb Mortal Wkly Rep. 58(4): 85–90.

Centers for Disease Control and Prevention. 2012. Botulism from drinking prison-made illicit alcohol—Utah 2011. MMWR Morb Mortal Wkly Rep. 61(39): 782–4.

Centers for Disease, Control and Prevention. 2012. Fatal foodborne Clostridium perfringens illness at a state psychiatric hospital—Louisiana, 2010. MMWR Morb Mortal Wkly Rep. 61(32): 605–8.

Centers for Disease Control and Prevention. 2013. Multistate outbreak of *Campylobacter jejuni* infections associated with undercooked chicken livers—northeastern United States, 2012. MMWR Morb Mortal Wkly Rep. 62(44): 874–6.

Centers for Disease Control and Prevention. 2013. Outbreak of staphylococcal food poisoning from a military unit lunch party—United States, July 2012. MMWR Morb Mortal Wkly Rep. 62(50): 1026–8.

Centers for Disease, Control and Prevention. 2013. outbreaks of cyclosporiasis—United States, June–August 2013. MMWR Morb Mortal Wkly Rep. 62(43): 862.

Centers for Disease, Control, and Prevention. 2013. Outbreak of *Escherichia coli* O104:H4 infections associated with sprout consumption—Europe and North America, May–July 2011. MMWR Morb Mortal Wkly Rep. 62(50): 1029–31.

Centers for Disease, Control, and Prevention. 2014. Centers for Disease Control and Prevention. Surveillance for Foodborne Disease Outbreaks, United States, 2012, Annual Report. Atlanta, Georgia: US Department of Health and Human Services, CDC.

Centers for Disease, Control, and Prevention. 2014. Multistate Outbreak of Listeriosis Linked to Commercially Produced, Prepackaged Caramel Apples. Available from http://www.cdc.gov/listeria/outbreaks/caramel-apples-12-14/.

Collier, M.G., Y.E. Khudyakov, D. Selvage, M. Adams-Cameron, E. Epson, A. Cronquist, R.H. Jervis, K. Lamba, A.C. Kimura, R. Sowadsky, R. Hassan, S.Y. Park, E. Garza, A.J. Elliott, D.S. Rotstein, J. Beal, T. Kuntz, S.E. Lance, R. Dreisch, M.E. Wise, N.P. Nelson, A. Suryaprasad, J. Drobeniuc, S.D. Holmberg and F. Xu. 2014. Outbreak of hepatitis A in the USA associated with frozen pomegranate arils imported from Turkey: an epidemiological case study. Lancet Infect. Dis. 14(10): 976–81.

Dalton, C.B., C.C. Austin, J. Sobel, P.S. Hayes, W.F. Bibb, L.M. Graves, B. Swaminathan, M.E. Proctor and P.M. Griffin. 1997. An outbreak of gastroenteritis and fever due to *Listeria monocytogenes* in milk. N. Engl. J. Med. 336(2): 100–5.

Division of Viral Diseases, National Center for Immunization, Centers for Disease Control Respiratory Diseases, and Prevention. 2011. Updated norovirus outbreak management and disease prevention guidelines. MMWR Recomm. Rep. 60(RR-3): 1–18.

Donnenberg, M.S. and T.S. Whittam. 2001. Pathogenesis and evolution of virulence in enteropathogenic and enterohemorrhagic *Escherichia coli*. J. Clin. Invest. 107(5): 539–48.

Food and Drug Administration. Peanut Corporation of America January 2009. Available from http://www.fda.gov/downloads/AboutFDA/CentersOffices/OfficeofGlobalRegulatoryOperationsandPolicy/ORA/ORAElectronicReadingRoom/UCM109834.pdf.

Food and Drug Administration. 2012. Bad Bug Book, Foodborne Pathogenic Microorganisms and Natural Toxins. Second ed. Center for Food Safety and Applied Nutrition (CFSAN) of the Food and Drug Administration (FDA), U.S. Department of Health and Human Services. Available online at: http://www.fda.gov/downloads/Food/FoodborneIllnessContaminants/UCM297627.pdf.

Frank, C., D. Werber, J.P. Cramer, M. Askar, M. Faber, M. an der Heiden, H. Bernard, A. Fruth, R. Prager, A. Spode, M. Wadl, A. Zoufaly, S. Jordan, M.J. Kemper, P. Follin, L. Muller, L.A. King, B. Rosner, U. Buchholz, K. Stark and G. Krause. 2011. Epidemic profile of Shiga-toxin-producing *Escherichia coli* O104:H4 outbreak in Germany. N. Engl. J. Med. 365(19): 1771–80.

Glass, R.I., U.D. Parashar and M.K. Estes. 2009. Norovirus gastroenteritis. N. Engl. J. Med. 361(18): 1776–85.

Haendiges, J., M. Rock, R.A. Myers, E.W. Brown, P. Evans and N. Gonzalez-Escalona. 2014. Pandemic *Vibrio parahaemolyticus*, Maryland, USA, 2012. Emerg. Infect. Dis. 20(4): 718–20.

Hall, A.J., B.A. Lopman, D.C. Payne, M.M. Patel, P.A. Gastanaduy, J. Vinje and U.D. Parashar. 2013. Norovirus disease in the United States. Emerg. Infect. Dis. 19(8): 1198–205.

Hall, A.J., M.E. Wikswo, K. Pringle, L.H. Gould and U.D. Parashar. 2014. Vital signs: foodborne norovirus outbreaks—United States, 2009–2012. MMWR Morb Mortal Wkly Rep. 63(22): 491–5.

Hedican, E.B., C. Medus, J.M. Besser, B.A. Juni, B. Koziol, C. Taylor and K.E. Smith. 2009. Characteristics of O157 versus non-O157 Shiga toxin-producing *Escherichia coli* infections in Minnesota, 2000–2006. Clin. Infect. Dis. 49(3): 358–64.

Herwaldt, B.L. 2000. *Cyclospora cayetanensis*: a review, focusing on the outbreaks of cyclosporiasis in the 1990s. Clin. Infect. Dis. 31(4): 1040–57.

Hlady, W.G. and K.C. Klontz. 1996. The epidemiology of Vibrio infections in Florida, 1981–1993. J. Infect. Dis. 173(5): 1176–83.

Imanishi, M., A.E. Newton, A.R. Vieira, G. Gonzalez-Aviles, M.E. Kendall Scott, K. Manikonda, T.N. Maxwell, J.L. Halpin, M.M. Freeman, F. Medalla, T.L. Ayers, G. Derado, B.E. Mahon and E.D. Mintz. 2014. Typhoid fever acquired in the United States, 1999–2010: epidemiology, microbiology, and use of a space-time scan statistic for outbreak detection. Epidemiol. Infect. 1–12.

Leshem, E., M. Wikswo, L. Barclay, E. Brandt, W. Storm, E. Salehi, T. DeSalvo, T. Davis, A. Saupe, G. Dobbins, H.A. Booth, C. Biggs, K. Garman, A.M. Woron, U.D. Parashar, J. Vinje and A.J. Hall. 2013. Effects and clinical significance of GII.4 Sydney norovirus, United States, 2012–2013. Emerg. Infect. Dis. 19(8): 1231–8.

Letchumanan, V., K.G. Chan and L.H. Lee. 2014. *Vibrio parahaemolyticus*: a review on the pathogenesis, prevalence, and advance molecular identification techniques. Front Microbiol. 5: 705.

Loharikar, A., A. Newton, P. Rowley, C. Wheeler, T. Bruno, H. Barillas, J. Pruckler, L. Theobald, S. Lance, J.M. Brown, E.J. Barzilay, W. Arvelo, E. Mintz and R. Fagan. 2012. Typhoid fever outbreak associated with frozen mamey pulp imported from Guatemala to the western United States, 2010. Clin. Infect. Dis. 55(1): 61–6.

Lorber, B. 1997. Listeriosis. Clin. Infect. Dis. 24(1): 1–9; quiz 10-1.

Lynch, M.F., E.M. Blanton, S. Bulens et al. 2009. Typhoid fever in the United States, 1999–2006. JAMA 302(8): 859–65.

Mandal, B.K. and J. Brennand. 1988. Bacteraemia in salmonellosis: a 15 year retrospective study from a regional infectious diseases unit. BMJ 297(6658): 1242–3.

McCollum, J.T., A.B. Cronquist, B.J. Silk, K.A. Jackson, K.A. O'Connor, S. Cosgrove, J.P. Gossack, S.S. Parachini, N.S. Jain, P. Ettestad, M. Ibraheem, V. Cantu, M. Joshi, T. Duvernoy, N.W. Fogg, Jr., J.R. Gorny, K.M. Mogen, C. Spires, P. Teitell, L.A. Joseph, C.L. Tarr, M. Imanishi, K.P. Neil, R.V. Tauxe and B.E. Mahon. 2013. Multistate outbreak of listeriosis associated with cantaloupe. N. Engl. J. Med. 369(10): 944–53.

Melton-Celsa, A., K. Mohawk, L. Teel and A. O'Brien. 2012. Pathogenesis of Shiga-toxin producing *Escherichia coli*. Curr. Top Microbiol. Immunol. 357: 67–103.

Neil, K.P., G. Biggerstaff, J.K. MacDonald et al. 2012. A novel vehicle for transmission of *Escherichia coli* O157:H7 to humans: multistate outbreak of *E. coli* O157:H7 infections associated with consumption of ready-to-bake commercial prepackaged cookie dough—United States, 2009. Clin. Infect. Dis. 54(4): 511–8.

Nguyen, Y. and V. Sperandio. 2012. Enterohemorrhagic *E. coli* (EHEC) pathogenesis. Front Cell Infect. Microbiol. 2: 90.

Onderdonk, A. and W. Garrett. 2010. Gas gangrene and other *Clostridium*-associated diseases. In Principles and Practice of Infectious Diseases: Churchill Livingstone Elsevier.

Ooi, S.T. and B. Lorber. 2005. Gastroenteritis due to *Listeria monocytogenes*. Clin. Infect. Dis. 40(9): 1327–32.

Ortega, Y.R. and R. Sanchez. 2010. Update on *Cyclospora cayetanensis*, a food-borne and waterborne parasite. Clin. Microbiol. Rev. 23(1): 218–34.

Parsons, R., J. Gregory and D.L. Palmer. 1983. Salmonella infections of the abdominal aorta. Rev. Infect. Dis. 5(2): 227–31.

Poka, H. and T. Duke. 2003. In search of pigbel: gone or just forgotten in the highlands of Papua New Guinea? P N G Med. J. 46(3-4): 135–42.

Saphra, I. and J.W. Winter. 1957. Clinical manifestations of salmonellosis in man; an evaluation of 7779 human infections identified at the New York Salmonella Center. N. Engl. J. Med. 256(24): 1128–34.

Scallan, E., R.M. Hoekstra, F.J. Angulo, R.V. Tauxe, M.A. Widdowson, S.L. Roy, J.L. Jones and P.M. Griffin. 2011. Foodborne illness acquired in the United States—major pathogens. Emerg. Infect. Dis. 17(1): 7–15.

Shapiro, R.L., C. Hatheway and D.L. Swerdlow. 1998. Botulism in the United States: a clinical and epidemiologic review. Ann. Intern. Med. 129(3): 221–8.

Silva, J., D. Leite, M. Fernandes, C. Mena, P.A. Gibbs and P. Teixeira. 2011. *Campylobacter* spp. as a Foodborne Pathogen: A Review. Front Microbiol. 2: 200.

Sobel, J. 2005. Botulism. Clin. Infect. Dis. 41(8): 1167–73.

Sobel, J., N. Tucker, A. Sulka, J. McLaughlin and S. Maslanka. 2004. Foodborne botulism in the United States, 1990–2000. Emerg. Infect. Dis. 10(9): 1606–11.

CHAPTER 3

Foodborne Infectious Diseases in the International Traveler

Ursula Kelly[1] and *Shivanjali Shankaran*[1,*]

Introduction

Thirty five members of a tour group flew from Boston to Los Angeles. Of them, 5 developed symptoms of nausea and vomiting with a sixth experiencing multiple episodes of diarrhea. Patients walked throughout the airplane with one toilet needing to be closed due to soilage. Eventually, 15 members of the group and 8 non-tour group passengers were diagnosed with the same illness, with the culprit identified in the stool of sick tour group members and one other passenger. What is the likely cause of this illness (Hannah et al. 2008)?

The above is a report on the transmission of Norovirus on an airplane in 2008 and demonstrates the ease of transmission of gastrointestinal illnesses amongst travelers. Foodborne illnesses are a significant concern in both developing and developed nations and there is no debate that international tourism is increasing. Between 2010 and 2030 the rate of international tourists is predicted to increase on an average of 3.3% per year (Launay 2013). When 20–50% of those tourists develop travel-related diarrhea and other infections, the significance of the annual increase can

[1] Assistant Professor, Department of Internal Medicine, Division of Infectious Diseases, Eastern Virginia Medical School, 825 Fairfax Avenue, Ste 410, Norfolk, VA 23507, USA.
Email: kellyum@evms.edu
* Corresponding author: shankas@evms.edu

be assessed (The Centers for Disease Control and Prevention: Travelers' Diarrhea 2015). Recent data from the GeoSentinal Surveillance Network showed that many travelers returning from international trips suffered from vaccine preventable diseases (LaRocque et al. 2012). Knowledge of common potential health threats can hasten pre and post-travel care and counseling and decrease travel-related infections. This chapter discusses the food and water borne bacteria, viruses and parasites that international travelers most commonly encounter.

Common Gastrointestinal Infections in the Returning Traveler

Campylobacter jejuni is the commonest pathogen isolated from travelers returning with diarrhea. Data from the Foodborne Diseases Active Surveillance Network (FoodNet) from 2004–2009 showed that out of more than 8000 cases of enteric infections associated with travel, campylobacter diarrhea was reported in 42% of those afflicted (Kendall et al. 2012). Though it is the leading cause of diarrhea around the world, risk is highest for those who travel to areas with poor food and water sanitation. Risk factors include ingestion of contaminated poultry or raw/unpasteurized milk. Water contaminated by animals is also a substantial risk (Allen et al. 2013). As with most foodborne infections, diarrhea or dysentery along with abdominal pain, vomiting and fevers is common. Additionally, Guillain-Barré syndrome and reactive arthritis are well known post infectious complications. Patients with AIDS may have a longer or more severe course (CDC Campylobacteriosis 2015).

Diagnosis is typically established by growth of curved gram negative rods cultured at 42°C. Self-limiting infections need adequate maintenance of hydration and electrolytes. In more serious infections, antibiotics such as the fluoroquinolones or Azithromycin can decrease duration of symptoms and should be considered in immunocompromised travelers. Unfortunately, increasing antibiotic resistance is being encountered, particularly, with the use of fluoroquinolones. Macrolides are good alternatives, though resistance to these antibiotics is also slowly increasing.

The majority of Salmonella infections (88.3%) associated with international travel are caused by nontyphoidal serotypes (i.e., other than *S. typhi* and *S. paratyphi*) compared to *Salmonella typhi* (7.7%) and *S. paratyphi* (3.9%). According to FoodNet data, nontyphoidal Salmonella was the second commonest enteric infection diagnosed in travelers at 32% (after Campylobacter). The commonest destinations were Mexico, followed by India, Jamaica and the Dominican Republic (Kendall et al. 2012). *S. typhi* infections are most commonly associated with travel to South Asia followed by travel to other parts of Asia, Africa, Latin America and South America. Salmonellosis is contracted via the feco-oral route from food or water contaminated by feces. While animals are reservoirs of nontyphoidal Salmonella, man is the only source of *S. typhi*. Nontyphoidal salmonellosis typically presents with nausea, vomiting, diarrhea, abdominal pain and fever. Most people have a self-limiting infection; however, invasive disease may occur in the elderly or the immunocompromised host. Bacteremia can lead to distant seeding leading to osteomyelitis or aortitis. Typhoid fever caused by *S. typhi* has a longer incubation period with high grade fevers, "pea soup" diarrhea, headache and anorexia. Rose spots may be seen on the skin. Complications can include intestinal perforation and hemorrhage or even death. Untreated infection can last a month. Diagnosis depends on isolation of the Salmonella species from stool, and less commonly from blood. In typhoid fever, diagnosis is based on clinical suspicion as well as positive blood, stool or bone marrow cultures (Allen et al. 2013). Treatment for uncomplicated nontyphoidal salmonellosis is supportive. Antimicrobial treatment should be considered in the elderly, the immunocompromised, those with severe symptoms and those at risk of invasive disease. Fluoroquinolones such as Ciprofloxacin are currently treatment of choice, though there have been increasing reports of resistance (O'Donnell et al. 2014). Alternatives include third generation cephalosporins, Trimethoprim-Sulfamethoxazole or Azithromycin. Those with invasive disease need longer courses of treatment. In patients with AIDS, a 4 week long course may be required. If there is relapse, long term antibiotics are needed till there is immune reconstitution. Fluoroquinolones are also the first line of treatment for typhoid fever. However, increasing resistance means that third generation cephalosporins such as Ceftriaxone are often employed. Resistance profiles from cultured *S. typhi* should be used to guide therapy (CDC Salmonellosis 2015, CDC Typhoid 2015). Food and water safety are imperative for prevention of infection. Additionally, travelers to areas with high risk of typhoid should be offered typhoid vaccination. The oral live attenuated vaccine should be avoided in patients with immunosuppression. For frequent travelers, booster doses are needed after five years. Antibiotics should be avoided within 72 hours of the live vaccine. The parenteral polysaccharide vaccine can be used in patients with immunosuppression with booster needed after two years of initial vaccination. Despite vaccination, food and water hygiene remains the most important intervention for prevention of Salmonella infection.

Escherichia coli bacteria are a predominant part of the resident flora of the gastrointestinal tract and are generally harmless. However, there are six pathotypes grouped as the diarrheagenic *E. coli*, which are responsible for large numbers of food and water borne illnesses, many of which are global and travel related (CDC *E. coli* 2015). In 2012 the Infectious Disease

Society of America published that pathogenic *E. coli* were responsible for more than a third of cases of traveler's diarrhea (Hill et al. 2006). The two forms most studied in relation to travel are Enterotoxigenic *E. coli* (ETEC) which results clinically in a watery diarrhea and Shiga toxin-producing *E. coli* (STEC), also known as Enterohemorrhagic *E. coli* (EHEC) which produces a bloody diarrhea and is known as hemolytic uremic syndrome (HUS). Illness from STEC is most commonly associated with the serogroup O157:H7, however data is emerging from outbreaks in Germany and Europe associated with other serogroups, such as O104:H4 (Alexander et al. 2012). While this infection is of global epidemiology, a review of the literature of over 51 studies of traveler's diarrhea found it to be most associated with travel to Latin America, Africa, and India (Shah et al. 2009). Unpasteurized products and undercooked meat tend to be among the more frequently identified sources of infection and outbreaks are global. Diagnosis of STEC can be confirmed with identification in the microbiology lab, and if shiga-toxin positive the sample can be further tested for the O serogroup. Diagnosis of ETEC is primarily a clinical diagnosis as not all laboratories are equipped to distinguish the varying forms of pathogenic *E. coli* (CDC *E. coli* 2015). Treatment of ETEC as for the other most common causes of traveler's diarrhea mentioned above, is supportive care and, based on severity and antimicrobials. Differing from this is the treatment of STEC, where antibiotics are not recommended across the board due to the potential for increased risk of HUS, which already complicates roughly six percent of STEC infections (Nelson et al. 2011).

Shigella are gram-negative rods of which four species are known to be pathogenic to humans: *S. dysenteriae, S. flexneri, S. boydii,* and *S. sonnei.* They are responsible for a significant number of travel related gastrointestinal infections and in a recent study of returning US travelers, it was found to be the third most common organism cultured in those with gastrointestinal symptoms. They spread fecal-orally and require a low concentration of inoculum for infection, which contributes possibly to the estimated 80–165 million cases/year seen worldwide (Brunette 2014). Although this is a global organism, the areas of highest travel-acquired risk include North Africa, Southeast Asia, India, and Oceana. They are invasive pathogens with a complex pathogenic mechanism which generally leads to bloody or mucoid diarrhea with fever (Hill et al. 2006). Symptoms begin 12–96 hours after infection and will last on an average for 4–7 days (Brunette 2014). Diagnosis is confirmed with stool culture and treatment is indicated with fluoroquinolones or azithromycin (Hill et al. 2006). While no vaccine exists currently, work is being done on an ETEC/Shigella vaccine with the hope of reducing the global burden of infection (Walker 2015).

Norovirus, of the family Caliciviridae, is the leading cause of gastroenteritis worldwide, with almost 20% of cases believed to be caused by it (Ahmed et al. 2014). Of the six designated genogroups, GI (Norwalk virus) and GII are believed to be responsible for most human illnesses, with Genogroup II genotype 4 being the leading cause of outbreaks worldwide. The mode of transmission is from person to person and feco-oral through contaminated food. Symptoms consist of nausea, vomiting and diarrhea with or without fever, which last typically 48–72 hours before complete resolution. Symptoms may be more severe, or even result in death, in the elderly (Mandell 2010). Treatment is largely supportive, with Reverse Transcriptase Polymerase Chain Reaction (RT-PCR) assays being widely used in diagnosis. Symptoms are more common in cooler months worldwide. Due to ease of transmission and very low infectious dose (18 particles), outbreaks are extremely common (CDC Norovirus 2015). Cruise ships are an ideal location for outbreaks with a large number of people interacting with each other in a small area. In 2014, the CDC's Vessel Sanitation Program (VSP) recorded 9 outbreaks of gastroenteritis on cruise ships. Seven of these were caused by Norovirus.

Travelers are recommended to follow food safety guidelines which include washing hands with soap and water, particularly after using the bathroom and prior to handling food. Those who are sick should avoid cooking food and attempts should be made to isolate the patient till 48 hours after resolution of symptoms (CDC Cruiselines 2015).

Haiti reported its first cholera outbreak in a century in 2010, a few months after a devastating earthquake. During the first year alone, almost half a million cases were reported with over 6631 deaths (CDC Haiti 2015). There were 129, 064 cases of cholera reported to the World Health Organization (WHO) in 2013 with 47% reported from Haiti and the Dominican Republic. Reports from Africa (43.6%) and Asia (9%) made up the remaining cases. This report showed that the number of cases reported worldwide is falling; however, it is likely that there is an underreporting of cases (WHO 2014). Travelers who follow safe food and water hygiene are unlikely to be at risk for infection with *Vibrio cholerae*. Travelers exposed to fecally contaminated food or water and health care workers volunteering during outbreaks are at risk for infection. Patients may be asymptomatic or mildly symptomatic. Severe cholera presents with large volume, "rice water" stool, profuse vomiting and severe dehydration and electrolyte imbalance which can lead to death if not treated promptly.

Diagnosis can be made by stool culture on selective media such as the thiosulfate-citrate-bile salts (TCBS) agar. Rehydration, either oral or parenteral, is the main treatment. Tetracyclines can decrease duration of illness in those with severe cholera. Macrolides and Fluoroquinolones are alternatives (CDC Cholera 2015). Prevention as always rests on washing hands, food and water hygiene and avoidance of undercooked/raw food. Currently, there are no vaccines available in the US.

Giardiasis and Amebiasis are common protozoan parasitic infections that can affect travelers. Both have a worldwide distribution, with travel to the tropics, subtropics and areas of poor sanitation having increased risk. Over 8% of travelers to the developing world seek care after their return (Allen et al. 2013). Data from the Geosentinal Surveillance System in the US showed that of over 13,000 diagnoses in returned travelers, giardiasis accounted for 3%. In the acute diarrhea group, giardia accounted for 13% of the diagnoses while amebiasis accounted for 4% (MMWR 2013). Longer duration of stay also increases risk of acquisition of both infections. Ingestion of fecally contaminated food, water and human contact are major routes of transmission, and sexual contact can also lead to spread. Both have a low infectious dose (< 100 CFU/ml). *Giardia lamblia* (or *G. intestinalis*) may cause a mild, short lived diarrhea or can lead to chronic diarrhea with weight loss, protein losing enteropathy, bloating, flatulence and cramps. Diarrhea may be foul-smelling and greasy, and post infectious irritable bowel syndrome may occur. While giardia has an incubation period of 7–10 days, it is slightly longer in Amebiasis (11–21 days). *Entamoeba histolytica* is the commonest causative organism, though *E. moshkovskii* can also cause human disease. Symptoms typically include bloody diarrhea, along with fever and abdominal pain. Similar to giardiasis, a majority of patients may be asymptomatic or have a self-limiting illness. Extra-intestinal disease (particularly in the liver) can occur rarely. Immunosuppression increases the risk of severe or invasive disease. Diagnosis is made by stool microscopy which may identify cysts or trophozoites, with multiple (three) stool samples increasing yield. Direct fluorescent antibody tests have high sensitivity as well. Ameba cysts or trophozoites can also be identified on stool microscopy; however, species differentiation cannot be made. As *E. dispar* is not pathogenic, stool or serum antigen assays are helpful with diagnosis of infection with *E. histolytica*. Serology can also be useful in diagnosis, though a positive test cannot differentiate between past or current infection. Treatment of symptomatic giardiasis consists of Metronidazole for five days or Tinidazole as a single dose. Other options include Nitazoxanide, Paromomycin, Furazolidone and Quinacrine. Intestinal amebiasis is also treated with Metronidazole (or tinidazole) and should be followed by a luminal agent such as Paromomycin to eradicate cysts within the lumen (CDC Giardiasis 2015, CDC Amebiasis 2015). As with all travel associated foodborne illnesses, prevention is paramount and consists of avoidance of contaminated food and water. As oral-anal or oral-genital sex is also a mode of transmission, patients should be counselled about safe sex practices.

Hepatitis A is a non-enveloped RNA virus (HAV) which infects hepatocytes causing both asymptomatic and symptomatic infections. It replicates in the hepatocytes, inciting inflammation, is excreted in bile into the stool and finally transmitted via fecal-oral route. This route of infection is promoted by the stability of the virus in the environment which may contribute to it being the second most common travel related infectious disease (Vaughn et al. 2014, Wu and Guo 2013). In general, HAV is endemic in developing countries, with the highest rates seen in Africa, India, and the Middle East, followed by Central and South America. A 2013 report established the risk of contracting HAV in travelers in medium to high endemic areas at 3/1000 to 20/1000 individuals per month traveled and is the second most common travel related infection seen (WHO Hepatitis A 2015). In developed countries community outbreaks are most often associated with shellfish, fruits and vegetables (Wu and Guo 2013). The course of the disease is variable, with the highest rates of morbidity and mortality in patients with underlying and chronic liver disease. The incubation period for HAV ranges from 10–50 days in which time the virus is replicating and can be transmitted. This is followed by a prodromal stage with fatigue, abdominal pain, nausea, vomiting, and dark urine with pale stools. Progression to the icteric stage is hallmarked by bilirubin levels greater than 20–40 mg/L and worsening of the prodromal symptoms. Of those that progress to fulminant liver failure increased mortality is associated with increased age, as well as underlying liver disease. Diagnosis is confirmed with anti-HAV IgM and treatment is generally supportive. Post-exposure prophylaxis is generally reserved for outbreaks, and the local/regional department of health guidelines can be consulted. The economic burden can be great due to the prolonged stages of the disease, with adults on an average missing 30 days of work (WHO Hepatitis A 2015). The Infectious Disease Society of America lists in its guidelines for travel medicine that vaccination against HAV should be considered for all travelers. Patients can be screened for immunity with HAV total antibody. If the patient has already been exposed to infection or vaccination, immunity is considered lifelong (Hill et al. 2006). A 2015 study evaluating the reasons why travelers did not receive vaccination against HAV found the leading cause was ignorance of existence of the vaccine (Liu et al. 2015). This highlights the benefits of pre-travel medical visits.

Hepatitis E (HEV) is the newest hepatitis virus discovered. First reported in 1980 and later found responsible for several Non A, Non B waterborne hepatitis outbreaks, dating back to the 1950s (Arankalle et al. 1994). It is similar to HAV in its routes of transmission (fecal-oral/foodborne) and the course of the disease. While it is considered endemic in developing countries, there are multiple examples of outbreaks in developed countries associated mostly with water and foodborne sources. It is specifically associated with raw or undercooked game meat, cattle and swine products and in 2008 was associated with a seafood related outbreak aboard a cruise ship (Arends et al. 2014, Perez-Garcia et al. 2014). A recent study from the Netherlands found rates of HEV in Western Europe to be underestimated and to be an impactful source of viral hepatitis (Koot et al. 2015). Similar to HAV, HEV is diagnosed with serum IgM/IgG studies. The clinical course is generally shorter than HAV, however there is a higher mortality rate in pregnant women, and interestingly it can become a chronic infection in immunocompromised patients (Perez-Garcia et al. 2014, Price 2014). A vaccine has been licensed in China, but is not yet available internationally (WHO HEV 2015).

Brucella species is the commonest zoonotic infection worldwide. It is most often acquired through consumption of unpasteurized dairy products (Dean et al. 2012a, Franco et al. 2007). It is thought that the process of making cheese can concentrate the Brucella organisms, where they can survive several months under the right conditions. Organ meats and fresh blood are a less frequent source of infection. After mucosal invasion by the bacteria it becomes an intracellular pathogen, with a pathogenetic and host response relationship which is quite complex. The incubation period is between 2 and 4 weeks followed by acute, subacute, or chronic infections. A 2012 review of human symptoms of brucellosis found that fever, arthralgia, myalgia, and back pain were among the most common symptoms reported, and a malodorous perspiration is known to be pathognomonic (Pappas et al. 2005, Dean et al. 2012b). Arthritis, sacroiilitis, and spondylitis are the most common complications of Brucella infection. However many organ systems can be involved, with CNS involvement and endocarditis being among the most serious with the highest mortality. Diagnosis is often made via blood culture. Though more invasive, bone marrow culture has a higher sensitivity due to accumulation of the bacteria in the reticuloendothelial system. Blood culture plates should be held for a prolonged period of time and the microbiology lab should be notified of the possibility of this diagnosis. Serologic diagnosis is a useful tool but cross-reactivity and unknown significance in endemic regions must be considered (Pappas et al. 2005). As travel patterns broaden so does the reach of infectious diseases and brucella is no exception. Traditionally locations most associated with Brucella have been the Mediterranean basin, South America, and the Eastern Europe and Asia. However, international tourism, among other contributing factors, has brought new areas of concern to the global map of brucellosis. According to the most recent WHO report on brucellosis, imported cases related to cheeses and other dairy products account for most of the acute cases of brucellosis seen in North America and Northern Europe. Combination therapy, with doxycycline/rifampin or doxycycline/streptomycin is generally employed for several weeks duration. Although relapse rates can be high (4–24%) mortality is quite low (no more than 2% of cases) (Franco et al. 2007).

Tapeworms of the genus *Diphyllobothrium* are among the largest intestinal parasites to infect humans. Consumption of infected raw fish and meats is the method of transmission to humans and with increasing global distribution of food products this infection is thought to have growth potential. It is also felt to be underdiagnosed as it is frequently asymptomatic and undetected for many years after infection (Kuchta et al. 2013). When symptoms do arise it is generally in association with intestinal obstruction and gall bladder disease, and long term parasite burden can cause vitamin B-12 deficiency and its sequelae (CDC Diphyllobothrium 2015). Human infections have been reported reliably in Asia, most frequently in Japan and Russia, with the coast of the Sea of Japan averaging 100 cases per year consistently and the coast of the Okhotsk Sea also considered endemic (Scholz et al. 2009). Diagnosis is made by standard stool exam for ova and parasites and is fortunately easy to treat with a onetime dose of praziquantel. Infection can be prevented by freezing the fish products properly eventually to be consumed raw as cold temperatures kill the larvae (CDC Diphyllobothrium 2015).

Immunocompromised Hosts

A 40 year old man with AIDS, who has received 6 cycles of chemotherapy for Lymphoma is placed on antiretroviral therapy (ART) and Co-trimoxazole for PCP prophylaxis. The patient is doing well on medications and travels to Thailand for several months. He continues with ART but stops prophylaxis. He returns to the clinic with a 2 month history of watery diarrhea and a 12 pound weight loss. He has between 2–5 watery bowel movements per day and is afebrile. His CD4 count is improved at 166 and there is no evidence of a recurrence of his Lymphoma. What could be the cause of his diarrhea (Egloff et al. 2001)?

Patients with Acquired Immunodeficiency Syndrome (AIDS, CD4 < 200) are known to be at risk for contracting opportunistic infection. They are also at risk for acquiring foodborne illnesses. In the above case report, the patient was diagnosed with Cyclospora induced diarrhea which responded to ART and Sulfamethoxazole-trimethoprim. The differential diagnosis in a patient such as this is quite large, so this discussion will be limited to protozoal infections such as *Cryptosporidium, Cyclospora, Microsporidium* and *Cystoisospora* (formerly *Isospora*). These four protozoan parasites are known to cause persistent diarrhea in people returning from international travel. They account for 3–6% of diarrhea in returning travelers (CDC Cryptosporidiosis 2015, Goodgame 2003).

Cryptosporidium hominis and *C. parvum* have a worldwide distribution. They are typically spread by the feco-oral route via contaminated food and water. Recreational exposure to water (swimming pools) can be a cause of outbreaks of cryptosporidiosis. They are resistant to chlorine as well as alcohol based hand sanitizers. In immunocompetent hosts, including patients with Human Immunodeficiency Virus (HIV) infection with CD4 count > 200, infection is typically self-limited. Symptoms usually consist of nausea, vomiting and diarrhea with low grade fevers or abdominal cramps. The incubation period is typically 7–10 days and symptoms usually resolve in 2–3 weeks. In patients with AIDS, diarrhea can persist for much longer causing wasting and electrolyte loss. Cryptosporidium is closely associated with AIDS cholangiopathy leading to elevated bilirubin and other liver function test abnormalities. Diagnosis is made by enzyme immunoassays (EIA), direct fluorescent antibody or modified acid fast (AFB) staining of stool. In the immunocompetent host, treatment is supportive, including replacement of fluids and electrolytes. When therapy is required, Nitazoxanide has been used with some benefit. In patients with AIDS, reconstitution of the immune system with the use of ART is the only proven treatment. Nitazoxanide addition to ART may provide some benefit in such patients. Prevention consists of appropriate hand washing with soap and water, avoidance of contaminated food and water and avoidance of recreational water spaces when symptomatic (i.e., diarrhea) (CDC Cryptosporidiosis 2015).

Cyclospora cayetanensis can cause traveler's diarrhea, with most cases usually reported after travel to tropical and sub-tropical countries such as Nepal, Guatemala and Peru. Unlike Cryptosporidium which can infect animals, humans are believed to be the only host for Cyclospora. Transmission is again through contaminated food and water. Presentation can include nausea, vomiting, watery diarrhea, flatulence, bloating, fatigue and anorexia as in the case above. Younger children and patients with AIDS may have more severe symptoms and may suffer from significant weight loss. The duration of the disease is typically about 3 weeks though immunosuppressed hosts tend to have more chronic symptoms. Diagnosis depends on identification of oocysts in stool. Modified AFB stains and fluorescent staining may also be used. Trimethoprim-Sulfamethoxazole for 7–10 days is the treatment of choice. Treatment may have to be extended for longer periods in patients with AIDS (CDC Cyclosporiasis 2015, Ortega and Sanchez 2010).

Many Microsporidia have been identified as causing human disease with *Enterocytozoon bieneus* and *Encephalitozoon intestinalis* being the commonest implicated species. These first came into prominence when they were identified as causes of diarrhea and wasting in AIDS patients.

Transmission is via the feco-oral route and microsporidia can affect humans as well as animals. Though uncommon, microsporidia have been known to cause chronic diarrhea in travelers, with symptoms of watery diarrhea and nausea generally lasting 4–6 weeks. Extra-intestinal infection may be seen, particularly in the eyes in the form of keratitis (Didier and Weiss 2011). There have been reports of cases amongst immunocompetent travelers returning from the tropics and the subtropics (Wichro et al. 2005). In a series of four travelers diagnosed with microsporidiosis in Spain, diarrhea resolved in the three immunocompetent patients, but had a more chronic course in the patient with HIV (Lopez-Velez et al. 2006). Modified trichome stain can be used for diagnosis. Albendazole is the treatment of choice, with shorter courses needed for immune competent individuals. In HIV+ patients, immune reconstitution is essential for cure. *E. bieneus* is less responsive to albendazole; Fumagillin has shown success, however, its significant bone marrow toxicity and lack of availability in the US limit its use (CDC Microsporidiosis 2015).

Cystoisospora belli (formerly *Isospora belli*) is common in the tropics and subtropics, and is spread via the feco-oral route by contaminated food and water. Typically associated with chronic, wasting diarrhea in severely immunocompromised patients (such as those with AIDS), there have been occasional reports of self-limiting diarrhea in travelers to countries in Asia and Africa (Perez-Ayala et al. 2011, Agnamey et al. 2010). Diagnosis is based on identifying oocysts in the stool, and may require special staining such as AFB or fluorescent techniques. Trimethoprim-Sulfamethoxazole is the drug of choice for treatment with longer courses needed for patients with significant immunosuppression. Ciprofloxacin is an alternative for treatment (CDC Cystoisospora 2015).

Prevention

Travelers may wish to consult a travel medicine physician at least 4–6 weeks prior to departure, especially if traveling to developing countries. Vaccinations based on the destination, are needed with underlying immunosuppression guiding which vaccines are safe. Live vaccines should be avoided in immunocompromised patients such as those with HIV infection or organ transplant recipients. Travelers with HIV and a CD4 count less than 200 are at a higher risk for opportunistic infections including the protozoan parasites described above. Data for travel in patients with other immunosuppressive conditions varies with regards to effectiveness of vaccines and risk of opportunistic infections. Hand washing and good food hygiene, including avoidance of raw produce or meat, tap water and unpasteurized dairy products are essential for all travelers. The use of bottled water, vigorous boiling or chemical treatments can help decrease risk of ingestion of contaminated water. It should be noted that Cryptosporidium and Cyclospora are resistant to treatment with chlorine. Those visiting developing countries should carry an appropriate antibiotic for self-treatment of traveler's diarrhea and should seek care if symptoms are worsening (CDC Immunocompromised 2015, UCSF 2009). Many foodborne illnesses may have a more protracted, atypical or severe course in patients with immunosuppression. Returning travelers with severe diarrhea, dysentery, chronic symptoms, abdominal pain or fever should seek immediate care.

Conclusion

In the age of increasing global connectedness, health care providers should be cognizant of the many food and water borne illnesses that may be imported after international travel. Knowledge of common causative pathogens, distribution of infections and varying presentations of gastrointestinal illnesses can help providers diagnose and contain them. Travel medicine clinics have the unique advantage of being able to reach at risk populations in a preemptive manner by providing counseling, vaccinations and strategies to maximize health while away from home. In particular, immunosuppressed patients may benefit from such pre-travel discussions.

References

Agnamey, P., Djamal Djeddi, Zahïr Oukachbi, Anne Totet and Christian P. Raccurt. 2010. *Cryptosporidium hominis* and *Isospora belli* Diarrhea in Travelers Returning From West Africa Journal of Travel Medicine. 17(2): 141–142.

Ahmed, S.M., A.J. Hall, A.E. Robinson, L. Verhoef, P. Premkumar, U.D. Parashar, M. Koopmans and B.A. Lopman. 2014. Global prevalence of norovirus in cases of gastroenteritis: a systematic review and meta-analysis. Lancet Infect. Dis. 14(8): 725–30.

Alexander, D.C., Weilong Hao, Matthew W. Gilmour, Sandra Zittermann, Alicia Sarabia, Roberto G. Melano, Analyn Peralta, Marina Lombos, Keisha Warren, Yuri Amatnieks, Evangeline Virey, Jennifer H. Ma, Frances B. Jamieson, Donald E. Low and Vanessa G. Allen. 2012. *Escherichia coli* O104:H4 Infections and International travel. EID. 18(3): 473–476.

Allen, G.P. Ross, G. Richard Olds, Allan W. Cripps, Jeremy J. Farrar and Donald P. McManus. 2013. Enteropathogens and chronic illness in returning travelers. N. Engl. J. Med. 368: 1817–25.

Arankalle, V.A., Mandeep Chadha, Sergei Tsarev, Suzanne Emerson, Arun Risbud, Kalyan Banerjee and Robert Purcell. 1994. Seroepidemiology of water borne hepatitis in India and evidence of a third enterically-transmitted hepatitis agent. Proc. Natl. Acad. Sci. 91: 3428–3432.

Arends, J.E., V. Ghisetti, W. Irving, H.R. Dalton, J. Izopet, A.I.M. Hoepelman and D. Salmon. 2014. Hepatitis E: An emerging infection in high income countries. J. Clin. Virol. 59: 81–88.

Brunette, GW. 2014. CDC Health Information for International Travel 2014: The Yellow Book. Elsevier.

Dean, A.S., Lisa Crump, Helena Greter, Esther Schelling and Jakob Zinsstag. 2012a. Global burden of human brucellosis: A systematic review of disease frequency. PLoS Negl Trop Dis. 6(10): e1865.

Dean, A.S., Lisa Crump, Helena Greter, Jan Hattendorf, Esther Schelling and Jakob Zinsstag. 2012b. Clinical manifestations of human brucellosis: A systematic review and meta-analysis. PLoS Negl. Trop. Dis. 6(12): e1929.

Didier, E.S. and Louis M. Weiss. 2011. Microsporidiosis: Not just in AIDS patients. Curr. Opin. Infect. Dis. 24(5): 490–495.

Egloff, N.,Thomas Oehler, Marco Rossi, Xuan M. Nguyen and Hansjakob Furrer. 2001. Chronic watery diarrhea due to co-infection with *Cryptosporidium* spp. and *Cyclospora cayetanensis* in a Swiss AIDS patient traveling in Thailand. Journal of Travel Medicine. 8(3): 143–145.

Franco, M.P., Maximilian Mulder, Robert H. Gilman and Henk L. Smits. 2007. Lancet Infect. Dis. 7: 775–86.

Goodgame, R. 2003. Emerging causes of traveler's diarrhea: Cryptosporidium, Cyclospora, Isospora, and Microsporidia. Current Infectious Disease Reports. 5(1): 66–73.

Hill, D.R., Charles D. Ericsson, Richard D. Pearson, Jay S. Keystone, David O. Freedman, Phyllis E. Kozarsky, Herbert L. DuPont, Frank J. Bia, Philip R. Fischer and Edward T. Ryan. 2006. The practice of travel medicine: guidelines by the infectious diseases society of America. CID. 43: 1499–1539.

http://hivinsite.ucsf.edu/InSite?page=kb-03-01-09#S2.2X. Accessed 1/25/15.

http://www.cdc.gov/dpdx/microsporidiosis/tx.html. Accessed 1/25/15.

http://www.cdc.gov/ecoli/general/index.html. Accessed 1/28/15.

http://www.cdc.gov/haiticholera/haiti_cholera.htm. Accessed 1/29/15.

http://www.cdc.gov/mmwr/preview/mmwrhtml/ss6203a1.htm. Accessed 1/25/15.

http://www.cdc.gov/nceh/vsp/cruiselines/norovirus_summary_doc.htm Accessed 1/20/15.

http://www.cdc.gov/ncidod/dbmd/diseaseinfo/travelersdiarrhea_g.htm. Accessed 2/19/15.

http://www.cdc.gov/norovirus/hcp/diagnosis-treatment.html. Accessed 1/20/15.

http://www.cdc.gov/parasites/cystoisospora/health_professionals/index.html. Accessed 1/25/15.

http://www.cdc.gov/parasites/diphyllobothrium/Accessed on 2/19/2015.

http://www.who.int/emc Accessed on 2/19/15.

http://www.who.int/mediacentre/factsheets/fs280/en/ Accessed 2/19/15.

http://wwwnc.cdc.gov/travel/yellowbook/2014/chapter-3-infectious-diseases-related-to-travel/campylobacteriosis. Accessed 1/27/15.

http://wwwnc.cdc.gov/travel/yellowbook/2014/chapter-3-infectious-diseases-related-to-travel/salmonellosis-nontyphoidal. Accessed 1/29/15.

http://wwwnc.cdc.gov/travel/yellowbook/2014/chapter-3-infectious-diseases-related-to-travel/typhoid-and-paratyphoid-fever. Accessed 1/29/15.

http://wwwnc.cdc.gov/travel/yellowbook/2014/chapter-3-infectious-diseases-related-to-travel/cholera. Accessed 1/29/15.

http://wwwnc.cdc.gov/travel/yellowbook/2014/chapter-3-infectious-diseases-related-to-travel/giardiasis. Accessed 1/26/15.

http://wwwnc.cdc.gov/travel/yellowbook/2014/chapter-3-infectious-diseases-related-to-travel/amebiasis. Accessed 1/26/15.

http://wwwnc.cdc.gov/travel/yellowbook/2014/chapter-3-infectious-diseases-related-to-travel/cryptosporidiosis Accessed 1/25/15.

http://wwwnc.cdc.gov/travel/yellowbook/2014/chapter-3-infectious-diseases-related-to-travel/cyclosporiasis Accessed 1/25/15.

http://wwwnc.cdc.gov/travel/yellowbook/2014/chapter-8-advising-travelers-with-specific-needs/immunocompromised-travelers. Accessed 1/25/15.

Kendall, M.E., Stacy Crim, Kathleen Fullerton, Pauline V. Han, Alicia B. Cronquist, Beletshachew Shiferaw, L. Amanda Ingram, Joshua Rounds, Eric D. Mintz and Barbara E. Mahon. 2012. Travel-associated enteric infections diagnosed after return to the United States. Foodborne Diseases Active Surveillance Network (FoodNet), 2004–2009 CID. 54(S5): S480–7.

Kirking, Hannah L., Jennifer Cortes, Sherry Burrer, Aron J. Hall, Nicole J. Cohen, Harvey Lipman, Curi Kim, Elizabeth R. Daly and Daniel B. Fishbein. 2010. Likely Transmission of norovirus on an Airplane, October 2008. CID. 50(9): 1216–1221.

Koot, H., B.M. Hogema, M. Koot, M. Molier and H.L. Zaaijer. 2015. Frequent hepatitis E in the Netherlands without traveling or immunosuppression. J. Clin. Virol. 62: 38–40.

Kuchta, R., Jan Brabec, Petra Kubackova and Tomas Scholz. 2013. Tapeworm Diphyllobothrium dendriticum (Cestoda)—Neglected or Emerging Human Parasite? PLoS Negl. Trop. Dis. 7(12): e2535.

LaRocque, R.C., Sowmya R. Rao, Jennifer Lee, Vernon Ansdell, Johnnie A. Yates, Brian S. Schwartz, Mark Knouse, John Cahill, Stefan Hagmann, Joseph Vinetz, Bradley A. Connor, Jeffery A. Goad, Alawode Oladele, Salvador Alvarez, William Stauffer, Patricia Walker, Phyllis Kozarsky, Carlos Franco-Paredes, Roberta Dismukes, Jessica Rosen, Noreen A. Hynes, Frederique Jacquerioz, Susan McLellan, DeVon Hale, Theresa Sofarelli, David Schoenfeld, Nina Marano, Gary Brunette, Emily S. Jentes, Emad Yanni, Mark J. Sotir, Edward T. Ryan, and the Global TravEpiNet Consortium. 2012. Global TravEpiNet: A National Consortium of Clinics Providing Care to International Travelers—Analysis of Demographic Characteristics, Travel Destinations, and pretravel healthcare of high-risk US International Travelers, 2009–2011. CID. 54(4): 455–62.

Launay, O. 2013. Vaccination of immunocompromised travelers: need for specific recommendations! Journal of Travel Medicine. 20(5): 275–277.

Liu, S.J., U. Sharapov and M. Klevens. 2015. Patient Awareness of need for Hepatitis A Vaccination (Prophylaxis) before international travel. J. Travel Med. Jan. 24. [Epub ahead of print].

Lopez-Velez, R., M. Carmen Turrientes, Carla Garron, Pedro Montilla, Raquel Navajas, Soledad Fenoy and Carmen del Aguila. 2006. Microsporidiosis in Travelers with Diarrhea from the Tropics. Journal of Travel Medicine. 6(4): Article first published online: 26 JUL 2006.

Mandell, G.L. 2010. Norovirus and Other Caliciviruses. Mandell, Douglas, and Bennett's Principles and Practice of Infectious Diseases (Chapter 175) Philadelphia, PA. Churchill Livingstone Elsevier.

Nelson, J.M., P.M. Griffin, T.F. Jones, K.E. Smith and E. Scallan. 2011. Antimicrobial and antimotility agent use in persons with shiga toxin-producing *Escherichia coli* O157 infection in FoodNet Sites. CID. 52(9): 1130.

O'Donnell, A.T., Antonio R. Vieira, Jennifer Y. Huang, Jean Whichard, Dana Cole and Beth E. Karp. 2014. Quinolone-Resistant Salmonella enterica serotype enteritidis infections associated with international Travel. CID. 59(9): e139–e14.

Ortega, Y.R. and Roxana Sanchez. 2010. Update on *Cyclospora cayetanensis*, a food-borne and waterborne parasite. Clin. Microbiol. Rev. 23(1): 218–234.

Pappas, G., Nikolaos Akritidis, Mile Bosilkovski and Epameinondas Tsianos. 2005. N. Engl. J. Med. 52: 2325–36.

Perez-Ayala, Begona Monge-Maíllo, Marta Díaz-Menendez, Francesca Norman, Jose A. Perez-Molina and Rogelio Lopez-Velez. 2011. A self-limited travelers' diarrhea by isospora belli in a patient with dengue infection. Journal of Travel Medicine. 18(3).

Pérez-Garcia, M.T., Beatriz Suay and María Luisa Mateos-Lindemann. 2014. Infect Genet Evol. 22: 40–59.

Price, J. 2014. An update on hepatitis B, D, and E viruses. Top Antivir. Med. 21(5): 157–63.

Scholz, T., Hector H. Garcia, Roman Kuchta and Barbara Wicht. 2009. Update on the Human Broad Tapeworm (Genus *Diphyllobothrium*), Including clinical relevance. Clin. Microbiol. Rev. 22(1): 146–160.

Shah, N., Herbert L. DuPont and David J. Ramsey. 2009. Global etiology of travelers' diarrhea: systematic review from 1973 to the present. Am. J. Trop. Med. Hyg. 80(4): 609–614.

Vaughan, G., Livia Maria Goncalves Rossi, Joseph C. Forbi, Vanessa S. de Paula, Michael A. Purdy, Guoliang Xia and Yury E. Khudyakov. 2014. Hepatitis A virus: Host interactions, molecular epidemiology and evolution. Infect. Genet. Evol. 21: 227–43.

Walker, R.I. 2015. An assessment of enterotoxigenic *Escherichia coli* and *Shigella* vaccine candidates for infants and children. Vaccine. 33(8): 954–965.

Wichro, E., David Hoelzl, Robert Krause, George Bertha, Franz Reinthaler and Christoph Wenicsh. 2005. Microsporidiosis in travel associated chronic diarrhea in immune competent patients. Am. J. Trop. Med. Hyg. 73: 285–287.

World Health Organization Weekly epidemiological Record No. 31 2014, 89, 345–356.

Wu, D. and C.Y. Guo. 2013. Epidemiology and prevention of hepatitis A in travelers. J. Travel Med. 20(6): 394–9.

CHAPTER 4

Outbreak Investigation in Foodborne Infectious Diseases—Successes and Limitations

Tara Nicolette[1] *and Nancy M. Khardori*[1,*]

Introduction

A long with diagnostic testing and treatment in managing foodborne infectious diseases, outbreak investigation is the most critical and difficult task. There is always an overabundance of factors that have to be taken into account when investigating an outbreak. The possible sources of food contamination covers a huge range that starts at the farm and leads to the table. There are many handlers of food and sources of contamination that can occur during all the stages of food preparation (from slaughter to serving). We also have to take into consideration location of the outbreak, epidemiology of the contaminant and the area in which the outbreak occurred and the virulence of the pathogen, among other things. The essential questions that are asked during an outbreak are: where did it originate, what is transmitting it, who is becoming infected, why did it occur, and how do we control and manage it.

[1] Division of Infectious Diseases, Department of Internal Medicine, Eastern Virginia medical School, Norfolk , Virginia, USA.
* Corresponding author: KhardoNM@EVMS.EDU

Definition

The first hurdle to overcome in foodborne infectious disease outbreaks is the definition of what constitutes an outbreak. The CDC guidelines for Confirmation of Foodborne-Disease Outbreaks states that a foodborne-disease outbreak (FBDO) is defined as, "an incident in which two or more persons experience a similar illness resulting from the ingestion of a common food." Botulism used to be an exception to the rule due to the severity of life threatening illness that follows infection, but that was changed in 2009 to contain the same stipulation of requiring two cases in order to constitute an outbreak (CDC; MMWR 2000). However, the diagnosis of outbreak differs among institutions contained within the United States, not to mention the definitions utilized by other countries.

There are many different investigating agencies which can become, or may need to be, involved during a given outbreak. Depending on the size and location of the outbreak: local, state and federal government agencies many need to be involved. Often, local offices become involved in an outbreak initially and it may expand to involve other agencies as needed, depending on the size and scope of the outbreak. Local outbreaks are usually investigated by public health officials in just one city or county health department. If the outbreak crosses into other cities or counties then the state health department would likely handle the investigation. The state health department often works with the state department of agriculture and with federal food safety agencies (detailed below).

The main federal government branches that respond to outbreaks are: the Centers for Disease Control (CDC), United States Department of Agriculture (USDA), FSIS (Food Safety and Inspection, branch of the USDA), and the Food and Drug Administration (FDA). If the outbreak is felt to be related to covert activities or there is a concern that it may cross international borders then the World Health Organization (WHO) may also be involved.

The Food Safety and Inspection Service (FSIS) is the branch within the United States Department of Agriculture (USDA) that "works with other government agencies and private sector organizations to ensure that it is able to respond quickly and effectively to an attack on the food supply, major disease outbreak, or other disaster affecting the national food infrastructure." FSIS is also the main branch of the USDA that monitors and controls the safety of the meat, poultry and processed egg products for public consumption. They are first responders to any incidents involving the food supply during natural disasters or terrorist attacks where the food supply could be compromised. This is obviously a very important part of outbreak prevention and management.

The USDA has many regulations on the handling and care of meats that have become standardized to help prevent the outbreaks from occurring at all. And, as important as this branch is, physicians and lay people do not usually utilize this system for reporting. The reporting to the FSIS is usually done by health inspectors or other workers in meat handling facilities if there is risk for contamination or improper handling being observed. FSIS and the FDA work closely together to oversee U.S. food safety and regulate the food industry with inspection and enforcement. In the case of an outbreak of foodborne illness, one or all of these agencies, work to find out why it occurred, take steps to control it, and look for ways to prevent future outbreaks. They also have the important job of announcing possible food recalls if needed.

The CDC usually leads investigations of "widespread outbreaks" or "those that affect many states at once". The CDC is most often involved in outbreaks that involve large numbers of people or severe or unusual illness and obviously their involvement/assistance may be requested by the state in which the outbreak has occurred. And, the CDC routinely collaborates with the other federal food safety agencies (ex. FDA, FSIS) during an outbreak investigation (CDC: Investigating Outbreaks 2013). The coordinated function of various agencies is responsible for the day to day safety of our food supply as well as prompt response to outbreaks.

Limitations

The most important limitation in the process of foodborne outbreak investigation is the delay in reporting. FoodNet is one of the main foodborne infectious disease reporting systems for the US. FoodNet includes 10 state health departments as well as information from the CDC, FDA, USDA, and the US Department of Agriculture, but only accounts for about 15% of the US population. The current states include Connecticut, Georgia, Maryland, Minnesota, New Mexico, Oregon, Tennessee, and selected counties in California, Colorado, and New York. Which means that it excludes information from 40 states as well as the other 85% of the US population. The same can be said of their reportable foodborne illness causative agents, which includes: **laboratory-confirmed cases** of infection caused by *Campylobacter, Listeria, Salmonella,* Shiga toxin-producing *Escherichia coli* (STEC) O157, STEC non-O157, *Shigella, Vibrio,* and *Yersinia*; and two parasites: *Cryptosporidium* and *Cyclospora* [FoodNet Surveillance: Active Laboratory Surveillance, CDC website]. Their statistics do seem to correlate fairly well with the national averages, according to their study (Jones et al. 2004).

One of the retrospective studies done through FoodNet, helps highlight the barriers to the successes of foodborne outbreak investigations. The large majority of the cases (71%) had no identified etiologic organisms. The ones that did have an identifiable cause usually had larger case sizes; > 10 cases involved rather than the standard 2 or more cases as defined by the CDC. They found that 66% of the outbreaks were associated with restaurants, 9% were associated with catered events and 7% were identified as being associated with private homes or gatherings (Jones et al. 2004). However, there is a natural skew with the preponderance to report restaurant outbreaks as opposed to an outbreak in one's own home. Also, the likelihood of transmitting the agent to others is less likely in the home as opposed to a restaurant.

In 9% of the outbreaks investigated, a "food service worker" was the source of contamination. Their role in the contamination was specified as: "handling by an infected person or carrier of pathogen" (21 of 44 outbreaks), "inadequate cleaning...leads to contamination of vehicle" (12 of 44 outbreaks), or food service workers "allowing foods to remain at room or warm outdoor temperature for several hours" (18 of 38 outbreaks).

Only a small amount, 36%, of cases had a laboratory-confirmed etiology. Part of this was likely due to outbreak case sizes but much of it was also due to lack of specimens. In 50% of cases no small laboratory studies were sent or stool samples collected during the outbreak. In other cases where stool studies were sent, testing was done only for select bacteria and parasites. This makes for a very difficult situation for websites like FoodNet and other reporting sites to provide meaningful epidemiologic information. However, initially with a small number of people affected and quick resolution and recovery of the people involved likely, some physicians may be less inclined to gather and send stool samples (which need to be collected properly for higher yield) unless the symptoms last longer than 24–48 hour. The study also demonstrated that more of the cases were investigated by the state health department and identified an etiologic agent for the outbreak, as opposed to those investigated only by local or county health departments. This also is fairly expected due to the fact that the CDC and state health departments are more likely to be involved in cases in which more people and more extensive areas are affected, with more potential specimens for testing, and more success in finding an etiology (Jones et al. 2004). "Not surprisingly, outbreaks in which ≥ 1 stool specimen was collected were significantly more likely to have their etiology identified than outbreaks in which no stool specimens were collected."

Reporting done through FoodNet is actually fairly easy. They check the reports of the seven bacteria and two parasites (as noted above) that are followed through their system about every 6 months from the 10 state health departments involved in the FoodNet surveillance system. The obvious limitation of this is the fact that there must be a laboratory confirmed diagnosis of the illness to be reported from the health department. This creates a fairly large bias due to the limited number of stool lab samples sent to state health departments by physicians. Also, if the state health department is not involved in the FoodNet surveillance and an outbreak is not reported to the CDC, it becomes very easy to miss the smaller outbreaks which affect health and health care costs.

Every year, FoodNet releases an annual summary report of all the information collected through the surveillance system which includes the incidence and number of identified pathogens in reported foodborne outbreaks (from both local and state health departments). They also release the Morbidity and Mortality Weekly Report (MMWR) and are able to project the estimated annual cost to the health care system and the estimated number of cases that are not reported for every 1 case that is reported.

So, where do physicians fit in all this? We are probably a large part of the limitation to the surveillance system. All the different government branches have different response systems, different ways of reporting outbreaks and different surveillance methods. Most of these are outlined in large documents that are easily accessible on their respective websites. The fact that reporting would have to be done physically by the physician, or by their office staff, already makes it less likely to be reported. Another limitation is the reporting of laboratory results if the specimens are not sent to a state laboratory, or not reported to the state laboratory by a local laboratory and are not reported to the CDC through their website. In that case a large number of outbreaks could be occurring without any record and epidemiology.

For example, foodborne disease outbreaks (FBDOs) should be reported to the Foodborne and Diarrheal Diseases Branch at CDC on Form 52.13, Investigation of a Foodborne Outbreak. This takes time and effort on the part of busy physicians and their office staff. Also, with physicians, much of their diagnosis comes from clinical judgment, especially when it comes to diarrheal and vomiting illnesses. One of the tools that can be helpful in diagnosis is the timeline. Different viruses and bacteria have different incubation periods which can help differentiate the types of food involved or the environment that harbors them. But many of these resolve in a short period of time, making physicians less likely to report the illness or collect stool or other samples for laboratory testing.

The second barrier is the collection of specimens. Many of the foodborne illnesses are diarrheal illnesses which call for stool collection for testing. An adequate number of stool samples should be collected and these need to be sent to the laboratory quickly for better yield. For example, testing for viruses (specifically norovirus) can require up to six specimens for accurate testing. Another issue is the potential for infection of staff through handling of samples, and education for appropriate use of personal protective equipment (PPE) is essential. There is a MMWR on the CDC website to help with choosing which lab tests to order called: "Recommendations for Collection of Laboratory Specimens Associated with Outbreaks of Gastroenteritis."

Outbreaks can be particularly difficult to track these days due to extensive travel (both international and within the US), large numbers of people at sporting events and endurance events. There was a recent "mud run" near Las Vegas, Nevada in which 22 cases of Campylobacter infection were identified and eventually traced back to the race. It was a long-distance obstacle adventure race, which has become more and more popular recently. During these races there are certain obstacles, many of which include some sort of mud pit that is created by the race officials, or through the trampling of the dirt by the racers themselves. Many of the courses of such races cut through farmland or other similar fields which are usually inhabited by both wild and domestic animals. Through the animal feces becoming intermixed with the mud in these runs, there is exposure to many different bacteria and increased risk of infection. Most of the Campylobacter infections that occur are associated with raw poultry, or other food items that have been contaminated by raw or undercooked poultry (CDC; MMWR 2014).

Cruise ships are also a growing venue for foodborne infectious disease, the most common cause being Norovirus. A study published in 2011 showed how behavior of passengers changed during a Norovirus outbreak. Noroviruses are highly transmittable, highly virulent and easily spread from person to person, through contaminated food and water, and contact with contaminated surfaces or objects. An outbreak on a cruise ship is defined as ≥ 3.0% of a cruise ship's population presenting to the ship's infirmary with an acute gastrointestinal illness (associated with diarrhea with or without vomiting). The viral infection in this case was likely brought on to the ship by an already ill passenger. With the high virulence and environmental stability of Noroviruses, it spread easily. The most prominent risk factor was a sick cabin mate, associated with a 3-fold increase for developing acute gastroenteritis. Some of the passengers demonstrated improved hand hygiene, using either hand-washing or alcohol based products, when they were informed of the outbreak. Although alcohol based sanitizers are popular and may help to kill the virus, hand-washing with soap and water is still the gold standard. The study also showed that although passengers were sick with a GI illness there was little self-isolation. There was a decrease in the amount of activities they were involved in but the illness did not confine them to their rooms, making these passengers the carriers of the virus to other unaffected passengers (Wikswo et al. 2011).

In addition to more people traveling there is transfer of food products across international borders. The US shares a border with Mexico. There is significant trade of food products between countries and one of those is unpasteurized cheese products, like queso fresco. People are increasingly choosing unpasteurized milk and cheese products due to health fads and misinformation, increasing health risks associated with them. There are multiple myths put forth by uneducated people: the pasteurizing process destroys the nutritional value of milk/cheese, removes the vitamins, causes lactose intolerance while unpasteurized products do not, or raw dairy products are not dangerous and are better for your health (FDA 2011). When in fact, the contrary is true. Unpasteurized dairy products are 150 times more likely to cause a foodborne infection than pasteurized dairy products. Listeria, and a host of other bacteria that are killed by the pasteurization process, have the potential for infecting susceptible individuals and causing foodborne outbreaks.

Recently there have been multiple reports of Listeria outbreaks in the news. The most recent reports have been on contaminated apples used to make commercially produced pre-packaged caramel apples (January 2015), quesito casero (fresh curd cheese) (2014), and contaminated cantaloupes (2011). Listeria, although it has low infectivity, is an important foodborne illness due to its high mortality rate (20–25%), hardiness, and extensive sources of contamination. Listeria can grow in a wide range of temperatures, from the temperature of a refrigerator (39.2°F or 4°C) to the body's internal temperature (37°C or 98.6°F). Listeria can contaminate: "hot dogs, deli meats, pasteurized or unpasteurized milk, cheeses (particularly soft-ripened cheeses like feta, Brie, Camembert, blue-veined, or Mexican-style *queso blanco*), raw and cooked poultry, raw meats, ice cream, raw vegetables, and raw and smoked fish." Ingestion of Listeria bacteria, and subsequent infection, causes listeriosis which is a serious illness that can cause meningitis, encephalitis, sepsis, fetal loss, and brain abscess. Listeria affects pregnant women, newborns, immunocompromised people, and the elderly (people over the age of 65 most commonly).

One of the main limitations found during the investigation of pre-packaged caramel apples was that the testing was not advanced enough to differentiate the strains of Listeria found in the suspected contaminated food products and the strains cultured from ill patients suspected of ingesting contaminated products. The laboratory analysis of the strains of Listeria were so similar, they were not able to be differentiated by pulsed-field gel electrophoresis (PFGE), one of the main laboratory tests used during outbreaks. This makes for a difficult situation for both the manufacturer of the food products and the investigating teams, in this case the CDC and FDA. As the Listeria from the apples collected from the patients during illness and suspected contaminated food products could not be matched definitively, assumptions had to be made, warning notices were put in places, and significant recalls were done to protect the public. Investigators had also to depend on a very thorough history, and the patient's memory of ingesting the food products, to help discern the culprit of the outbreak (CDC: Listeria [Listeriosis] 2015).

Water

Another potential source of contamination and subsequent outbreak, is recreational water such as public pools, lakes, etc. In one case study they review the causes of outbreaks associated with recreational water during 1971–2000. The article defined "recreational water" as being: "swimming and wading pools, thermal and other natural springs, fresh and marine waters, water parks, interactive fountains, and other venues where water contact may take place." Some of the water was classified as treated and others as untreated (fresh and marine waters), both of which can become contaminated. At least two people showed similar symptoms after contact with water and epidemiological evidence that strongly indicated the source of water in question was the most likely origin of infection.

Gastroenteritis is by far the most common manifestation of illness associated with these outbreaks, although dermatitis is the second most common manifestation in waterborne outbreaks (usually caused by Pseudomonas aeruginosa). As with foodborne outbreaks, the fecal-oral route is the most common means of infection but this can also be said of waterborne outbreaks. Some of the other illnesses that were less common are: conjunctivitis (usually caused by adenovirus), pharyngitis, aseptic meningitis, keratitis, otitis externa, hepatitis, and typhoid fever. The most common venues where outbreaks took place are lakes and streams (fresh water untreated), and pools and public fountains (treated).

The most common reasons found for outbreaks were fecal accidents (36% treated, 31% untreated) and ill bathers; followed by diaper-age children, high bather density, heavy bather use, and bather crowding being the other likely reasons for contamination and outbreak. Some of the other factors that were not as common but were cited, were related to sewage and animal runoff (some during heavy rains). The reasons for outbreaks were traced back to human error in a large number of the filtered water cases due to inadequate maintenance, meaning unsatisfactory disinfection or filtration/treatment of the water which was being used.

Of the reported cases, a bacteria or protozoan (most commonly Shigella and Cryptosporidium) was most commonly identified and was as the cause of illness (79% of cases). Cryptosporidium (38%) being most common in diaper-age children, *E. coli* O15:H7 and Shigella (18%) being the most common with fecal accidents, and Norovirus (8%) and Giardia (5%) making up the other agents in cases where etiology was established. Seventy-five of the 259 outbreaks (29%) resulted in one or more hospital admissions. The diagnoses were found to be: "shigellosis (32%), cryptosporidiosis (24%), *E. coli* O157:H7 gastroenteritis (18%), leptospirosis (14%), and shigellosis (~ 4 cases)."

As with most of the sources of outbreaks, the core limitations to sporadic waterborne illness epidemiology is lack of information, reporting, and education. Fortunately, reporting of waterborne outbreaks has risen steadily over the past 30 years. While there were only about 10 reported outbreaks in 1971–1975, this number rose to over 90 reported cases of outbreak in 1996–2000 (Craun et al. 2005).

A case in point, one bacteria that highlights both the successes and limitations in investigation of foodborne illness is typhoid fever. As the times have changed, so too have some of the causes of foodborne outbreaks. Historically, typhoid has been a major cause of illness and foodborne disease, and even though it has decreased greatly in scale it is still an important source of foodborne illness. In an article published online by FoodSafetyNews.com in 2012, number 1 and 2 of the top ten deadliest foodborne outbreaks in US history were caused by typhoid. One of the most famous people associated with foodborne outbreaks is known as Typhoid Mary, real name Mary Mallon, who transmitted and infected people through her cooking. From 1920–1940 typhoid fever was the culprit of 70% of waterborne outbreaks, by 1941–1960 it had dropped to 22%, by 1971–1990 it fell to 11%, and now it causes scattered sporadic outbreaks (Craun et al. 2006). "In 1900, the incidence of typhoid fever was approximately 100 per 100,000 population; by 1920, it had decreased to 33.8, and by 1950, to 1.7 (CDC; MMWR 1999)." Once a major health threat, it has effectively been eradicated from the US in large part due to improved food handling standards.

Typhoid fever causes gastrointestinal upset (nausea, vomiting, diarrhea), high fevers (> 102°F), and often requires hospitalization. The causative agent is a bacterium called *Salmonella typhi* and is contracted when ingesting food or water contaminated with fecal material of an infected person. The illness can last up to 4 weeks if not treated. The morbidity and mortality is usually caused by severe dehydration, internal hemorrhaging, ileus formation, and septicemia.

Typhoid is still a major cause of illness in developing countries and up to 75% of cases in the United States are acquired through international travel. Past successes in the control of typhoid have been through vaccination in international travelers. Other than typhoid and Hepatitis A, there are no other available vaccines for prevention of foodborne illnesses. The vaccines are recommended if people are going to travel internationally to endemic areas for long periods of time. Even with the vaccine, there can still be outbreaks that can cause major morbidity and mortality (CDC: Typhoid fever 2013).

In 2013, a cafeteria worker at Purdue University was found to be have typhoid fever on the university campus. He had recently travelled overseas and brought the illness back with him. This did not result in an outbreak of typhoid on campus, even though he was a food handler, in large part due to the development of sanitary food handling conditions. He had been wearing gloves and hand washing is a large component of food service these days, effectively halting the spread of typhoid. He was also acutely infected with *Salmonella typhi* and was not an asymptomatic carrier like "Typhoid Mary." An asymptomatic carrier state creates a more difficult situation for investigation of foodborne outbreak by making diagnosis more difficult, as people do not show symptoms, and the spread of disease is more likely.

The most common cause of typhoid fever outbreaks in the United States remain food related. The most recent outbreak associated with typhoid fever was in 2010 in California and Nevada, with a total of 9 affected persons and no deaths. The source of contamination was traced back to an imported food product known as: Goya brand Frozen Mamey Fruit Pulp. Mamey is a popular tropical fruit native to Central and South America. The fruit pulp can be purchased in grocery stores throughout the United States (CDC: Salmonella 2010).

Conclusion

On the whole, foodborne illness in the United States has decreased significantly over time. Much of the decrease can be attributed to improvements in sanitation, refrigeration of food products and improved handling of food by producers, restaurant workers, and consumers. But foodborne outbreaks and illnesses remain difficult to trace and control, thanks to substandard sanitation in areas, mishandling of food products, shipping of food products internationally and limited reporting of suspected outbreaks. There is still much progress to be made in improving methods of tracking, reporting, and education that are vital in ensuring food safety.

Keywords: Foodborne illness, outbreak, Center for Disease Control (CDC), FoodNet, recreational water, cruise ship, typhoid, listeria

References

Author unknown. 2010, August 25. Multistate Outbreak of Human Typhoid Fever Infections Associated with Frozen Mamey Fruit Pulp (Final Update). CDC: Salmonella, 2010 outbreaks. Retrieved from http://www.cdc.gov.

Author unknown. 2011 Revision. Grade "A": Pasteurized Milk Ordinance. Food and Drug Admin. Retrieved from http://www.fda.gov.

Author unknown. 2013, May 14. Typhoid Fever. CDC: National Center for Emerging and Zoonotic Infectious Diseases. Retrieved from http://www.cdc.gov.

Author unknown. 2013, November 12. Key Players in Foodborne Outbreak Response: Local, State, and Federal Agencies. CDC: Investigating Outbreaks. Retrieved from http://www.cdc.gov.

Author unknown. 2015, January 15. Multistate Outbreak of Listeriosis Linked to Commercially Produced, Prepackaged Caramel Apples. CDC: Listeria (Listeriosis). Retrieved from http://www.cdc.gov.

CDC. Achievements in Public Health, 1900–1999: Safer and Healthier Foods. MMWR. 1999: 48(40): 905–913.

CDC. MMWR surveillance summaries: Appendix B, Guidelines for Confirmation of Foodborne-Disease Outbreaks. MMWR. 2000: 49(SS01): 54–62.

CDC. Outbreak of Campylobacteriosis Associated with a Long-Distance Obstacle Adventure Race—Nevada, October 2012. MMWR. 2014: 63(17): 375–378.

Craun, G.F., R.L. Calderon and M.F. Craun. 2005. Outbreaks associated with recreational water in the US. Int. J. Environ. Heal. R. 15(4): 243–262.

Craun, M.F., G.F. Craun, R.L. Calderon and M.J. Beach. 2006. Waterborne outbreaks reported in the United States. J. Water Health. 4(Suppl. 2): 19–30.

Jones, T.F., B. Imhoff, M. Samuel, P. Mshar, K.G. McCombs, M. Hawkins, V. Deneen, M. Cambridge and S.J. Olsen. 2004. Limitations to successful investigation and reporting of foodborne outbreaks: an analysis of foodborne disease outbreaks in foodNet catchment Areas, 1998–1999. Clin. Infect. Dis. 38(Suppl. 3): S297–302.

Wikswo, M.E., J. Cortes, A.J. Hall, G. Vaughan, C. Howard, N. Gregoricus and E.H. Cramer. 2011. Disease transmission and passenger behaviors during a high morbidity norovirus outbreak on a cruise ship, January 2009. Clin. Infect. Dis. 52: 1116–1122.

CHAPTER 5

Sources and Vehicles of Foodborne Infectious Diseases

Hariharan Regunath[1] and *William Salzer*[2,*]

Introduction

Foodborne infectious diseases cause significant morbidity despite the advances in food processing technology and medical care. In 2011, Centers for Diseases Control (CDC) estimated 48 million foodborne illnesses occur each year in the United States (US), of which 9.4 million episodes were from confirmed pathogens [5.5 million (59%) by viruses, 3.6 million (39%) by bacteria, and 0.2 million (2%) by parasites] (Scallan, Hoekstra et al. 2011, Scallan, Griffin et al. 2011). More recent data published by the Foodborne Diseases Active Surveillance Network (FoodNet) that monitors laboratory confirmed foodborne infections caused by the nine most common pathogens in 10 states in the US (15% of US population), reported 19,056 infections, 4,200 hospitalizations, and 80 deaths in 2013 (Crim et al. 2014).

The global burden of foodborne illness, outside the United States was largely unknown until the World Health Organization (WHO) Initiative to estimate the Global Burden of Foodborne Diseases, formed a Foodborne Disease Burden Epidemiology Reference Group (FERG)

[1] Fellow, Division of Infectious Diseases, Department of Medicine, University of Missouri, Columbia, MO 65212.
[2] Professor and Division Head, Division of Infectious Diseases, Department of Medicine, University of Missouri, Columbia, MO 65212.
* Corresponding author: salzerw@health.missouri.edu

to guide the generation of estimates of the global burden of foodborne diseases (Kuchenmüller et al. 2009). Global Foodborne Infections Network (GFIN) is the section of WHO collaborating with the CDC, European CDC, OzFoodNet (Australian surveillance system modelled from FoodNet) and other internationally recognized public health organizations to provide guidance and expertise for laboratory based surveillance, control of major foodborne illness and containment of antimicrobial resistance in foodborne pathogens (Global Foodborne Infections Network 2014, Green and Fitzsimmons 2013).

Outbreak investigations are the primary means of identification of culprit foods and the infectious pathogens transmitted by them. At least fifteen new food sources were identified as causes of outbreaks since 2006 (Painter et al. 2009, Painter et al. 2013, Braden and Tauxe 2013). Despite public health surveillance and outbreak investigations, the true incidence is often underestimated because a significant number of patients fail to seek medical attention and the rest are not recognized by physicians. Strict study criteria for outbreak investigations and non-availability of diagnostic tests in all health care settings, also contribute to this underreporting of cases (Gould et al. 2013).

The focus of this chapter is to review and discuss the various sources and vehicles of foodborne infectious diseases. Modes of transmission will be discussed as relevant and pertinent. Individual diseases and outbreaks are being discussed in other chapters.

Definitions of Food Categories

The sources and vehicles that transmit infectious pathogens have been classified to enable better analysis of outbreaks and planning of preventive measures (Painter et al. 2009, Gould et al. 2013, Pires et al. 2012, Painter et al. 2013, Braden and Tauxe 2013). The following definitions have been developed by the CDC to classify categories of foods implicated as sources or vehicles in foodborne disease outbreaks.

Ingredient: A basic component of a food item that does not contain any other food items and belongs to only one commodity.

Simple food: Food consisting of only one ingredient, e.g., Green leafy vegetables or eggs.

Complex Food: Food that contains two or more ingredients.

Commodity groups: Hierarchal groups of categories containing related commodities. E.g., beef and pork are grouped under commodity group "meat". Likewise broader hierarchal commodity groups can be plant or animal origin, processed or unprocessed foods, etc. Adapted from Painter et al. is Figure 1, displaying the hierarchy of food commodities.

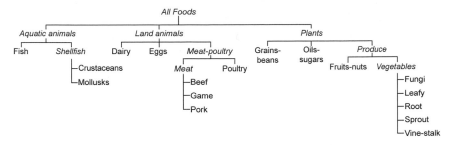

Figure 1. Hierarchy of food commodities. *Italicized* words are commodity groups (Painter et al. 2013).

Microbiology

Bacterial infections outnumber viruses and parasites as the most commonly identified cause of foodborne outbreaks, laboratory confirmed cases of nine bacterial pathogens are being monitored by the FoodNet in the United States. Similar networks in Europe and Australia provide data from their respective jurisdictions. The major bacterial pathogens reported in the most recent data from USA is listed in Table 1, in the decreasing order of their incidence rates. Viral and parasitic infections are listed in Table 2. A study aimed at food source attribution by reviewing the line listings of outbreaks from the CDC (1995–2005) and European Union Food Safety Authority (zoonotic outbreaks reports for 2004 and 2005), provided data from outbreaks across the globe from 1988–2007. The results of this study reported Salmonella (46.9%), Norovirus (13.5%) and *E. coli* (9.5%) as the identified causes for almost 70% of the 4093 foodborne outbreaks worldwide, but this is probably a skewed dataset because most of the outbreaks are from the United States followed by EU and Australia, and only to a minimal extent from other countries (Greig and Ravel 2009).

Food Commodities and Implicated Pathogens

The Foodborne Diseases Outbreak Surveillance System of the CDC receives outbreak data on the agents, foods and settings responsible for foodborne outbreaks from state health departments in the United States. An online platform, the Foodborne Outbreak Online Database (FOOD) is user friendly

Table 1. Number of cases and incidence of culture confirmed major bacterial agents of foodborne illness and their sources/vehicles from the CDC, USA (Crim et al. 2014, Nygren et al. 2013, McCollum et al. 2013).

Bacterial Pathogen	No. of cases in 2013 (Incidence in cases per 100,000)	Sources/Vehicles
Salmonella	7277 (15.19)	Multiple animals, eggs, poultry, undercooked ground meat, dairy products and fresh produce contaminated with animal waste, peanut butter.
Campylobacter	6621 (13.82)	Undercooked Poultry, contact with or contamination with wild or domestic animal waste, water, milk.
Shigella	2309 (4.82)	Poor hygienic practices in food handling, multiple food handlers in restaurants, institutionalized settings and day care centers.
STEC Non O157	561 (1.17)	Cattle and young calves, Fresh produce contaminated by manure from cattle, undercooked ground beef, water systems.
STEC O157	552 (1.15)	
Vibrio	242 (0.51)	Raw/uncooked shellfish, contaminated water.
Yersinia	171 (0.36)	Raw pork products, milk and spring water.
Listeria	123 (0.26)	Refrigerated cold cut or deli meat, cantaloupe, caramel apple, contaminated Mexican style cheese.

to search, access and download descriptive summaries (year, state, place of food consumption and causative agent) of outbreak data (Control et al. 2014).

An outbreak is defined as two or more cases of similar illness resulting from ingestion of a common food. Among outbreak investigations between 1998 and 2008, 58% had reported a specific food commodity as a source, of which poultry (19%), fish (19%) and beef (12%) were the most common sources (Gould et al. 2013). Less than half (42%) of these foods could be assigned to one of the 17 predefined commodities (Figure 1), of which poultry (17%), leafy vegetables (13%), beef (12%), and fruits/nuts (11%) were the commodities associated with the most outbreak-related illnesses, after assignment (Gould et al. 2013).

Overall, animal foods are more commonly implicated as a source than plant foods, complex foods are more common sources than simple foods and some pathogens are associated with multiple commodities further adding to complexity (Greig and Ravel 2009).

Table 2. Sources and vehicles of viral and parasitic infections reported from CDC (MacKenzie et al. 1995, Crim et al. 2014, Scallan, Hoekstra et al. 2011, Fankhauser et al. 2002, Lopez et al. 2001, Notes from the field: use of electronic messaging and the news media to increase case finding during a Cyclospora outbreak—Iowa, July 2013, Wheeler et al. 2005, Hutin et al. 1999).

Pathogen		Total No. of cases per year (percentage proven to be foodborne for viruses; incidence in 2013 per 100,000 for parasites)	Sources/Vehicles
Viruses	Norovirus	5,461,731 (26%)	Shellfish, clams, oysters, fomites containing the vomitus of infected persons, food contamination from infected water and food handlers especially in cruise ships, nursing homes, schools, etc.
	Hepatitis A virus	1566 (7%)	Raw or uncooked shellfish, oysters; contaminated vegetables—green onions and frozen strawberries.
	Astrovirus	15433 (< 1%)	Less extensively studied, but in general sources/ vehicles are similar to norovirus.
	Rotavirus	15433 (< 1%)	
	Sapovirus	15433 (< 1%)	
Parasites	Cryptosporidium	1,186 (2.48)	Contaminated drinking water and food from infected food handler.
	Cyclospora	14 (0.03)	Fruits (Guatemala raspberries) and raw vegetables (mesclun lettuce, basil, and snow peas.

In the following sections, we review the seventeen commodities reported in outbreaks primarily from the USA (some from EU, Australia and other countries) and the common pathogens transmitted. Most common causative agents for illnesses or hospitalization and their associated commodities varied with time period of surveillance studies and other specific risk factors involved. As the focus of this chapter is on the sources and vehicles of foodborne illness, details of individual disease manifestations is beyond the scope of this chapter. Likewise chemical toxins as foodborne outbreaks including biological toxins (ciguatera, paralytic shellfish, mycotoxins, scombroid, etc.) will not be discussed.

Animal Food

Land animal foods are classified as dairy products, eggs, meat and poultry, whereas aquatic animal foods are the sea foods. Of these, meat, poultry and seafood can be categorized into one group as "muscle foods" (Sofos et al. 2013). Generally, muscle tissue as such is considered sterile, but contamination with a multitude of bacteria and fungi occurs on their surface, during processes like animal slaughtering, dressing, catching fish and also from the environment (human hands during handling and preparation, equipment surfaces during production and processing, soil and water contamination, etc.). Whereas the number and diversity of bacteria causing infection is large, the common bacterial pathogens of concern for outbreaks have been only a noteworthy list, but nevertheless new or emerging pathogens are being added to the list (Sofos 2008). The common pathogens are *E. coli* (including *E. coli* O157:H7), *C. perfringens, Salmonella, Campylobacter* and *L. monocytogenes* for land muscle foods, and *Vibrio* spp., *L. monocytogenes* and *Salmonella* for sea foods (Sofos et al. 2013).

Meat: Red meat is beef, pork or meat of game. The slaughtering and carcass dressing procedure especially hide removal or skinning and evisceration result in contamination of the meat from the gut flora (Rhoades et al. 2009).

Ground beef has been identified as the leading cause of *E. coli* O157:H7 infections, and CDC surveillance data for 2009–2010, reported this pathogen-commodity pair as one of the leading causes of most outbreak related hospitalizations (Rangel et al. 2005; Two multistate outbreaks of Shiga toxin-producing *Escherichia coli* infections linked to beef from a single slaughter facility—United States 2008, 2010, Nastasijevic et al. 2009, Laboratory-confirmed non-O157 Shiga toxin-producing *Escherichia coli*—Connecticut, 2000–2005, 2007, Lanier et al. 2009, Lanier et al. 2011). Asymptomatic enteric and hide carriage of various Verocytotoxigenic *E. coli* (VTEC) (O26, O45, O91, O103, O111, O121, O145 and O157) in cattle serves as a source for contamination of meat during abattoir processing especially skinning and evisceration (Rhoades et al. 2009, Laboratory-confirmed non-O157 Shiga toxin-producing *Escherichia coli*—Connecticut, 2000–2005, 2007, Manning et al. 2007). The colonization rates (stool > hides) are higher in adult cattle in feedlots than calves, and more so in warmer months. Although the VTEC do not multiply at temperatures below 7°C, they remain viable at this temperature for more than a week or longer at lower temperatures (Uyttendaele et al. 2001). As they are rapidly killed by heating, adequate cooking eliminates the risk of disease (Rhoades et al. 2009, Smith et al. 2001). A few outbreaks that reported beef as the source for

Shigella (*S. Sonnei*) infections, were probably from hand contamination from the food handlers because of the very low infective dose for this bacteria, and viability at refrigeration temperatures in raw beef (Warren et al. 2007). Beef, pork and game meat were rarely implicated as sources in *C. jejuni* outbreaks (Gould et al. 2013, Zhao et al. 2001).

Clostridium perfringens, a ubiquitous spore forming gram positive bacteria is a common pathogen transmitted by beef (49% of *C. perfringens* related outbreaks) (Bennett et al. 2013). Spores survive at extreme temperatures and resist most cleaning procedures, sporulation occurs after ingestion and vegetative forms produce an enterotoxin (type A is most common) that causes diarrhea without vomiting (Brynestad and Granum 2002). Most outbreaks have occurred from consumption of roasted meats (or reheated gravy) prepared in large quantities well ahead of time (in institutions, food service establishments like restaurants, deli, etc.) (Bennett et al. 2013). Ingestion of heat stable toxin produced by sporulating *B. cereus* and *S. aureus* have led to a few foodborne outbreaks, transmitted from beef. In contrast to illness from *C. perfringens*, emesis is a predominant feature of the toxin mediated illness from these bacteria. Although *B. cereus* was primarily associated with rice dishes, in an analysis restricted to outbreaks reporting a single likely implicated dish, beef was attributed in 22% of confirmed (2/9) *B. cereus* and 10% of confirmed (6/58) *S. aureus* cases. Likewise, pork was attributed in 22% of confirmed (2/9) *B. cereus* and 67% of confirmed (39/58) *S. aureus* cases (Bennett et al. 2013).

Non typhoidal *Salmonella* top the list of bacterial causes of outbreaks reported in the United States and are the second leading cause in the European Union (Laufer et al. 2014, European Food Safety Authority 2014). They also colonize herbivores, especially cattle (hides > stool) with a peak in warmer months (Rivera-Betancourt et al. 2004). Because of its potential to remain viable even at 5°C, retail raw meat can harbor and transmit infection (Zhao et al. 2001). In all outbreaks from Salmonella between 1973 and 2011 in the United States, ground beef and processed meat have been the most common sources of infection (Laufer et al. 2014). CDC Surveillance data for 2009–2010 reported Salmonella from pork as one of the pathogen commodity pairs associated with most deaths (Surveillance for foodborne disease outbreaks—United States, 2009–2010, 2013).

L. monocytogenes, unlike most pathogens multiplies at refrigeration temperatures and hence can reach dangerous levels even at 4°C, but is killed by pasteurization and adequate cooking. For the same reason hide contamination rates of *L. monocytogenes* have been reported to be higher in winter and spring (Guerini et al. 2007). Consumption of organ meat confers higher risk than muscle, as listeria concentrate and multiple in liver, spleen, kidney and lymph nodes (Ryser and Buchanan 2013). This organism was present in up to 3.5% of retail ground beef samples in a US survey (Samadpour et al. 2006). Ready to eat or delicatessen meat that are sliced in-store rather than manufacturer packaged products are ~ 7 times more likely to harbor listeria (Lianou and Sofos 2007).

Among viral pathogens, norovirus is the major pathogen that has been attributed to transmission from delicatessen meat, because of contamination with bare hands and fomites from an infected individual. Other viruses transmitted include hepatitis A and rotavirus (Moe 2009, Malek et al. 2009, Gould et al. 2013).

Trichinellosis is the most notable parasitic disease transmitted from meat (Wilson et al. 2015, Holzbauer et al. 2014, Hall et al. 2012, Kennedy et al. 2009). Of the 84 confirmed cases reported to the CDC (from 24 US states and District of Columbia) between 2008–2012, source attribution was most often game meat, 49% (41/84) from bear meat, 26% (22/84) from pork products (including wild boar, commercial pork, and home raised swine) and 2% (2/84) each from deer meat and ground beef with rest being unknown sources. Of the 51 cases that reported the manner of cooking, 47% (24/51) reported consuming raw or undercooked meat and sausages made from game meat. Outbreaks are more commonly seen during peak hunting seasons, spring and fall. Freezing in most cases kills the organism, but some strains are resistant to freezing, hence adequate and complete cooking (71°C by non-microwave methods) is essential to prevent infection (Holzbauer et al. 2014).

Poultry: Poultry includes chicken and turkey meat, and all over the world these are the most commonly consumed meat second only to pork (Sofos et al. 2013). Poultry was the most commonly implicated single commodity among foodborne disease outbreaks reported to the CDC (Gould et al. 2013). The steps involved in poultry processing begins with stunning (renders them unconscious), followed serially by killing and bleeding, scalding (immersion in hot water), picking (removal of feathers), evisceration and chilling (Sofos et al. 2013). The last three processes, especially evisceration, are the most likely causes of surface contamination of retail poultry meat (Baker et al. 1987).

Salmonella and poultry, is the pathogen-commodity pair responsible for 145 outbreaks reported to the CDC between 1998 and 2008 (Gould et al. 2013). Stool carriage in infected birds serves as the primary source. Poultry acquire *Salmonella* by various routes—vertical (trans-ovarian to progeny or eggs), horizontal (direct contact from infected birds) or from the stool contaminated environment (water, feed, etc.). Vertical transmission results in amplification of horizontal transmission because of large number of birds living in close quarters (Park et al. 2008). Prevalence and bacterial counts in retail poultry meat depends on the type of retail store, storage temperature and poultry company (integrated large scale vs. non-integrated small scale). Due to widespread use of antimicrobials in poultry farms, drug resistance is prevalent in *Salmonella* and this confers public health risks (Donado-Godoy et al. 2012, Donado-Godoy et al. 2014).

Poultry consumption is a major risk factor for *C. jejuni* infection, as rates of contamination of retail raw poultry have been as high as 100% in the past (Allos 2001, Gould et al. 2013, Butzler 2004, Baker et al. 1987). Campylobacteriosis is a leading cause of foodborne gastroenteritis worldwide, and the leading cause of outbreaks in the EU (European Food Safety Authority 2014, Allos 2001). Important sequelae to *Campylobacter* associated diarrhea include reactive arthritis and Guillian Barre Syndrome (GBS). Apart from cross contamination from fecal material during evisceration of poultry, breach of hygienic practices in food handling have been implicated as the chief means of acquisition of *Campylobacter* infection. Likewise cross contamination of other foods (Salads, fruits, vegetables) sharing the same surface for cutting, handling and processing, can occur. Undercooked poultry confers highest risk to transmission; adequate and thorough cooking kills it effectively. Effective carcass decontamination procedures have been adopted in developed countries, resulting in a decline in the number of outbreaks in some parts of the developed world (e.g., Sweden) (Butzler 2004, Acheson and Allos 2001). In the United States, improved hygienic practices on farms and especially poultry slaughterhouses contributed to a declining trend from 1996 to 2005, however, the trend seems to increase gradually since 2006 (Incidence and trends of infection with pathogens transmitted commonly through food—foodborne diseases active surveillance network, 10 U.S. sites, 1996–2012, 2013).

Between 1998 and 1999, a multistate (22 states) outbreak of listeriosis occurred in the USA and the implicated source was opened and unopened packages of turkey franks from one facility of a major manufacturer (Mead et al. 2006). In the following years, delicatessen turkey meat slices were implicated in two other multistate (11 states in 2000, 9 states in 2002) outbreaks in the United States (Gottlieb et al. 2006, Olsen et al. 2005).

Other pathogens reported to be transmitted by poultry in outbreaks include toxin producing bacteria *C. perfringens*, *B. cereus* and *S. aureus*. Among viral illnesses, Norovirus is the only viral agent with source attribution to poultry, which is also primarily from unhygienic practices of food handling and contaminated water. Few outbreaks have reported Giardiasis from poultry.

Eggs: Salmonella is the major pathogen associated with eggs. *S. typhimurium* is most common in Australia and *S. enteritidis* and other nontyphoidal salmonella are more common in USA and EU (Threlfall et al. 2014, Li et al. 2013). Salmonella can contaminate the interior or exterior of the eggs from colonized flocks at a rate ranging from 0.3% to 1% of eggs (Threlfall et al. 2014). In the United States, of the 183 outbreaks from Salmonella between 1973 and 2001, 28 outbreaks involved foods containing eggs (Chittick et al. 2006). Raw egg products including mayonnaise, home-made ice cream, pre-prepared egg salad, eggnog and even scrambled eggs have been implicated in multiple outbreaks in Australia and USA. A never proven hypothesis is that Salmonella can penetrate egg shell because of its porosity to permit gas exchange for the developing embryo (Li et al. 2013).

Dairy: Dairy was the second most common source for infectious illnesses reported to the CDC from outbreaks between 1998 and 2008 (Painter et al. 2013). Milk from various sources (cattle, goat and sheep) and its products (butter, cheese, cream, etc.) constitute this category. Milk in the udder is considered to be sterile, but colonization occurs from milking (teat apex, equipment), containers, feeds, soil, water and other environments. Because of the nutrient composition and neutral pH characteristics, it serves as a good vehicle for survival and growth of bacteria, despite the presence of natural inhibitors (lactoferrin, lactoperoxidase, lysozyme, etc.).

Improper pasteurization or consumption of raw milk and their milk products has resulted in foodborne outbreaks from *Salmonella, Listeria, E. coli* O157:H7 and non O157 (e.g., O111), *Campylobacter, Brucella, M. bovis, Yersinia* and *C. burnetii* (Potter et al. 1984, *Escherichia coli* O157:H7 infection associated with drinking raw milk—Washington and Oregon, November–December 2005, 2007, *Escherichia coli* O157:H7 infections in children associated with raw milk and raw colostrum from cows—California, 2006, 2008, Salmonella typhimurium infection associated with raw milk and cheese consumption—Pennsylvania, 2007, 2007, Gillespie et al. 2003, Human tuberculosis caused by Mycobacterium bovis—New York City, 2001–2004, 2005). Cattle can serve as asymptomatic reservoirs for *C. burnetii, Listeria, M. avium* subspecies *paratuberculosis, Campylobacter, E. coli* and *S. enterica*. The frequency of contamination of pooled farm milk for foodborne pathogens is highest for *C. jejuni* (< 1–12.3%) followed by Salmonella (< 1–8.9%), Listeria (2.7–6.5%), Yersinia (1.2–.1%), and Shiga-Toxin producing *E. coli* (STEC) (< 1–3.8%) (Angulo et al. 2009). Prevalence rates for *C. burnetii* in samples of bulk tank milk from US dairy herds between 2001 and 2003, have been as high as > 94% by PCR (Kim et al. 2005). This bacteria causes abortion in sheep and disease of reproductive system in cattle. The current guidelines for temperature and time combinations for adequate pasteurization are based on the ability of the process to eliminate or kill *C. burnetii* (Angulo et al. 2009).

In 1985, dairy products were responsible for the largest outbreak of Salmonellosis in the United States, in which ~ 16,300 cases of confirmed Salmonellosis were associated with consumption of 2% milk from a common dairy farm. Even after extensive investigation by CDC and FDA, the cause could not be clearly ascertained, but conclusions by investigators were that a cross contamination of raw and pasteurized milk at a cross-connection pipe was probably the reason (Ryan et al. 1987). In addition, contaminated and unpasteurized cheese, ice cream, milk based powdered infant formula are other sources noted in outbreaks for Salmonellosis world-wide (Soler et al. 2008; Hennessy et al. 1996, Salmonella typhimurium infection associated with raw milk and cheese consumption—Pennsylvania 2007, Ratnam et al. 1999).

Among the outbreaks caused by VTEC (*E. coli non-O157*) from 2000–2010, dairy products are the most commonly attributed foods (Luna-Gierke et al. 2014).

Similarly, apart from contaminated pasteurized milk, consumption of Mexican/Hispanic style soft cheese (Queso fresco) or surface ripened variety has been a common vehicle for Listeriosis in USA, Canada and EU (Ryser and Buchanan 2013).

Milk products are also the most implicated vehicle for outbreaks from *Y. enterocolitica* in the USA. Outbreaks in New York in 1976 and 1981 were related to contaminated chocolate flavored milk and unhygienic reconstitution of powdered milk (Black et al. 1978, Shayegani et al. 1983). Yersinia have the ability to multiply at refrigeration temperatures by generation of cold shock proteins, which prepare the bacteria for cold acclimatization and continued growth (Robins-Browne 2013). An outbreak in 1982 in southern USA (Arkansas, Tennessee and Mississippi) involved pasteurized milk contaminated with swine manure (Tacket et al. 1984).

Improved animal health, hygienic milking practices and animal cleanliness are imperative measures for reducing the pathogen transmission rates, more importantly the avoidance of consumption of raw milk or unpasteurized dairy products.

Seafood: Finfish and Shellfish are the two broad categories of consumed seafood. Shellfish includes shrimp, crawfish, lobsters and other mollusks or crustaceans, which harbor bacteria representative of the source water bodies, harvesting methods, processing and food handling for marketing (Butt et al. 2004, Sofos et al. 2013). Contamination of water with human sewage is often implicated as the cause of outbreaks related to seafood.

In general finfish are implicated less often than shellfish, as the common illnesses from cooked finfish are more often from chemical toxicities (ciguatera or scombroid, chemicals/toxins in water) (Wallace et al. 1999).

Although an epidemiologic study that reviewed sea food related outbreaks (188 total outbreaks) reported to the CDC from 1973 to 2006, concluded that bacterial agents were the most common etiology (76%), followed by viral (21.3%) and parasitic agents (2.6%), more recent CDC surveillance data from 1998 to 2008 reported Norovirus related illnesses as the most common single etiology of outbreaks, followed by various bacteria and parasites (Gould et al. 2013, Wallace et al. 1999, Iwamoto et al. 2010). Norovirus which is primarily water borne, is the most common cause of non-bacterial acute gastroenteritis with a seasonal peak in winter months. Notably the bivalve mollusks store this virus and other microbes in large quantities as they feed by filtering large volume of water (Gentry et al. 2009, Robilotti et al. 2015). Hepatitis A virus is the other notable virus that has been transmitted from raw shellfish in outbreaks in the United States and Australia over the years (Desenclos et al. 1991, Bialek et al. 2007, Conaty et al. 2000).

Bacterial pathogens in seafood can come from indigenous flora of the marine/estuarine water bodies (*V. cholerae, V. parahemolyticus, V. vulnificus, L. monocytogenes, C. botulinum* and virulent strains of *Aeromonas hydrophila*), fecal contamination of water bodies with pathogenic enteric bacteria (*Salmonella* spp., pathogenic *E. coli, Shigella* spp., *Campylobacter* spp., and *Y. enterocolitica*) or contamination during processing (*C. perfringens, B. cereus, S. aureus* and *L. monocytogenes*) (Feldhusen 2000).

Salmonella and *C. botulinum* are the pathogenic bacteria most frequently identified from the final product of fish that is caught, whereas fish raised from recirculating aquaculture systems can harbor *S. aureus, L. monocytogenes* and *Vibrio* spp. (Sofos et al. 2013). Historically, Salmonellosis from seafood occurred as major outbreaks in Japan with cuttlefish (prolonged thawing at room temperature and inadequate boiling) and dried squid (contaminated well water in a processing plant) being the vehicles in 1988 and 1999. Seafood is generally not a common vehicle for Salmonella in the United States, but the few outbreaks implicating seafood as a vehicle involved salmon (2000), crab cake and lobster (2001), raw oysters (2003), shrimp and tuna (2004) and red snapper (2009) (Li et al. 2013). Pacific coast and Alaska are endemic for botulism. Home canned fermented seafood has been the major source of outbreaks of botulism (type E toxin) in Alaskan natives in 1990–2000. Other outbreaks included fresh fish (Palani) from Hawaii in 1990 and home salted fish in New Jersey in 1993 and 2007 (Johnson 2013).

Vibrio spp. are natural inhabitants of marine and estuarine environments and the species differ depending upon the salinity of the water body (e.g., *V. cholera* for fresh water, *V. parahemolyticus* and *V. vulnificus* for salt water). *V. parahemolyticus* has been a leading cause of seafood associated gastroenteritis, mainly from raw or undercooked shellfish in the US (McLaughlin et al. 2005, Newton et al. 2014). Irrespective of the species, warm temperatures favor multiplication of these organisms and hence most outbreaks have been noted in summer and fall. Global warming and changes in salinity of estuarine water bodies with rainfall also favor multiplication of pathogenic *Vibrio* spp. (Iwamoto et al. 2010, McLaughlin et al. 2005, Baker-Austin et al. 2010). Contaminated seafood and exposure of wounds to contaminated water are the chief modalities of disease acquisition with these bacteria.

High risk sea foods for Listeriosis include mollusks, raw or lightly preserved finfish, shellfish (salted marinated, fermented, cold smoked, etc.) or mildly heat processed fish products (Ryser and Buchanan 2013). In the United States, outbreaks from Giardiasis and Cryptosporidiosis have been attributed to bivalve mollusks as they concentrate the organisms and their cysts (Gould et al. 2013, Hohweyer et al. 2013). Because of this property, it has been proposed that certain mollusks can be used for biological monitoring of sediment contamination (Graczyk et al. 1999).

Adequate cooking eliminates the risk of pathogen acquisition, but as long as unhygienic practices in food handling prevail and consumption of minimally cooked or raw shellfish continues to be customary, the risk of an outbreak exists (Sofos et al. 2013).

Plant Food

Fresh Produce, both vegetables and fruits/nuts, can become contaminated at different points in their transport from harvest to consumption; from the soil (where fecal and other contamination from wild and domesticated animals and contaminated water can occur), during transport and processing (by farm handlers with poor hygiene and contaminated washings) to final preparation in the kitchen (unhygienic food handler, cross contamination from other infected foods sharing the cutting board or knives or utensils, unclean equipment, etc.) (Lynch et al. 2009). In the USA, among the 67,752 illnesses in the outbreaks between 1998 and 2008 that were assigned to a single commodity, leafy vegetables and fruits/nuts were associated in 13% and 11% respectively. Leafy vegetables were responsible for most illnesses (2.2 million [22%] either as a simple food or complex food) and were also the second most common cause of hospitalizations (Painter et al. 2013). So far there have been 25 outbreaks of fresh produce related outbreaks in

Australia between 2001 and 2006 (Angulo et al. 2008). In Europe up to 10% of foodborne outbreaks were linked to fresh produce in 2010 (The European Union summary report on trends and sources of zoonoses, zoonotic agents and foodborne outbreaks in 2010, 2012).

Major pathogens noted in these outbreaks included bacterial—VTEC (*E. coli* O157 and non O157), *Shigella, Campylobacter, Salmonella, S. aureus, C. perfringens* and *B. cereus*; viruses—Norovirus and Hepatitis A virus; and parasites—*Cyclospora* and *Cryptosporidium.*

Norovirus is the most common single pathogen responsible for most outbreaks (263 outbreaks and 8001 illnesses) attributed to fresh produce between 1998 and 2008 in the USA (Gould et al. 2013). Implicated fresh produce in Norovirus outbreaks were always linked to an unhygienic food handler or contaminated water (Robilotti et al. 2015). Surface contamination of fruits (grapes and berries) and leafy greens have been proven by various studies using molecular and genetic methods to detect norovirus in fresh produce, but their link to real outbreak situation remains to be ascertained (Baert et al. 2011, Tian et al. 2011, Robilotti et al. 2015). As human norovirus is not cultivable, various agents for surface decontamination (chlorine and sodium bicarbonate) have been explored in studies using animal models, i.e., murine norovirus, but no recommendations regarding the use of these agents have been made at this time. Adequate washing of fresh produce, hand hygiene of food handlers and disinfection of equipment surfaces are key measures necessary for prevention of contamination of fresh produce and resultant outbreaks (Bae et al. 2011, Robilotti et al. 2015).

Hepatitis A virus, because of its survival at refrigeration temperatures (4°C) including freezing, has resulted in a few outbreaks in USA from frozen fresh produce primarily imported from geographic areas where hepatitis A is endemic or prevalent. Contaminated frozen strawberries (1997, 1 outbreak) and green onions (2003, serial outbreaks) from northern Mexico were the classic outbreaks (Jaykus et al. 2013). More recently, frozen pomegranate arils from Turkey were implicated in a multistate outbreak of Hepatitis A virus in the United States in 2013 (Collier et al. 2014). Semi-dried tomatoes from Turkey have resulted in outbreaks in Europe and Australia (Céline et al. 2011, Donnan et al. 2012).

E. coli O157:H7 caused multiple outbreaks in the USA (1996 and 2006), the most notable ones were caused by contaminated fresh baby spinach and lettuce from California. Source tracing isolated the same strain from soil, water and animal feces (cattle and swine) in the produce cultivation area located on a ranch, indicating contamination from animals in the environment (Doyle and Erickson 2008). Other countries that reported major outbreaks from lettuce, alfalfa and other sprouts include—Japan

(1996, multiple outbreaks—9451 cases from contaminated white radish sprouts) and Europe (Meng et al. 2013). Fruit juice related outbreaks were noted in 1990s (mainly unpasteurized apple juice made from apples that came in contact with soil contaminated with animal feces) in the USA and apart from contamination, acid tolerance of *E. coli* O157:H7 at refrigeration temperature is suspected to be the reason (Besser et al. 1993).

Fresh produce has been implicated more often as a source for Salmonellosis within the past decade. Data from most worldwide outbreaks report the following foods as sources (country and year of outbreak mentioned in parenthesis following each food): Tomatoes (USA—1993, 2002, 2004, 2006; Canada—2006; Europe—2011), lettuce (Europe—2000, 2004, 2005), cantaloupes (USA and Canada—1991, 2002), papayas (USA—2011; Australia—2007), beans (USA—2000; Europe—2000, 2010) alfalfa sprouts (USA—1996, 2001, 2009; Canada—1999; Europe—1994), orange juice (USA—1995, 1999, 2000; Canada—1999; Australia—1999), mangoes (USA and Brazil—1999) and salad and salad base/bars (USA—1984; Canada—1991; Europe—1981, 2005) (Li et al. 2013). Likewise the only significant pathogen transmitted through nuts and dried seeds or cereals is also Salmonella.

When compared to other bacterial agents, *B. cereus* outbreaks have not been as frequent in the USA (Europe has more *B. cereus* outbreaks). Most *B. cereus* outbreaks between 1998 and 2008 in the USA (44/235 outbreaks causing 281/2050 illnesses) were associated with grains/beans (Gould et al. 2013). Because of the ubiquitous presence of Bacillus spores, spices and other seeds, especially rice can become contaminated during growth, harvest and other agricultural handling/processing. Up to 52.8% of the rice products from retail food stores have been shown contain *Bacillus* spores, which because of their hydrophobic characteristics (especially *B. cereus* more than other species) are difficult to remove by cleaning (Ankolekar et al. 2009). Unless the contaminated food is sufficiently heated (> 100°C), spores survive normal cooking temperatures, and when stored at room temperature germination occurs with multiplication and elaboration of toxin (emetic toxin causing emesis and enterotoxin causing diarrhea). Thus heat treated rice products like fried rice or pasta kept unrefrigerated for four hours can cause outbreaks (Granum and Lindbäck 2013).

Cyclosporiasis has been reported from a variety of fresh produce related outbreaks. Contaminated Guatemalan raspberries (USA—1995, 1996, 1997, 2000; Canada—1996, 1997, 1998, 1999, 2000), Basil from Mexico (USA—1999, 2005; Canada—2001), Salads and leafy herbs (EU—2000) and snow peas (USA—2004) were the vehicles noted from outbreaks (Herwaldt and Ackers 1997, Ortega and Sanchez 2010). Sporulation of the *Cyclospora* cysts is the key step for infectivity and this process takes 2 weeks under optimal lab conditions. This suggests that fresh produce must get contaminated with sporulated cysts to cause infection after consumption. But at this time other facts are not clear, about the mechanisms of transmissions, infective dose and seasonal variation (most cases occurred in summer in USA) (Ortega 2013).

Foodborne cryptosporidiosis is more often reported from the EU, especially the Nordic countries (Norway, Sweden and Denmark) than the United States and Australia, as fresh produce is more commonly consumed in these countries (Robertson and Chalmers 2013). Few outbreaks of cryptosporidiosis have been associated with apple cider made in (USA—1993, 1996, 2003), salad mixture (Sweden—2008), lettuce (USA—2009; Finland—2008), tomatoes (USA—2009), onions (USA 2009), Parsley (Sweden—2008), Carrots (EU—2005, 2009) and mung bean sprouts. Being a primary waterborne pathogen, contamination from infested water, environment (e.g., Feces from grazing animals contain a large amount of cysts) and infected and unhygienic food handlers are the modes of transmission. The cysts' ability for survival is notable because of resilience to chlorine or freezing, but they are readily inactivated by adequate heating (Duhain et al. 2012, Robertson and Chalmers 2013).

Conclusions and Future Perspectives

The importance of adequate cooking, cleanliness and hygeinic practices in handling, processing, packaging and dispensing methods of various foods cannot be overemphasized. Outbreak investigations provide the vital information needed for source attribution to specific foods, to enable appropriate steps to stop the spead of an outbreak and prevent future outbreaks. Advances in regulations for food processing, product screening and advanced pathogen detection methods such as whole genome sequencing (WGS) have raised further hope for generating results better and earlier than the present traditional methods that rely on time consuming culture methods (Salter 2014). Because of advances in field of clinical microbiology, many culture independent diagnostic tests (CIDT) that detect pathogen specific antigens or DNA are now available and an increasing number of clinical laboratories have adopted such methods for pathogen detection (Iwamoto et al. 2015). The advantages are rapid diagnosis, lower cost, the ability to test a wide array of pathogens at same time and less stringent requirements for specimen collection/processing. But, at present there are multiple brands for CIDT, the sensitivity of CIDT are not uniformly high enough for all potential pathogens causing outbreaks and the non-availability of culture isolates for further processing at public health laboratories have raised concerns for a negative impact on public health surveillance (Cronquist et al. 2012). Detailed documentation on the type of test used, reflex culturing of CIDT positive isolates and development of CIDTs for further strain characterization have been suggested as possible solutions (Iwamoto et al. 2015). Effective partnership among CIDT test developers, health care providers, clinical laboratories and regulatory agencies is also required.

Keywords: Outbreaks, foodborne illness, epidemic

References

Acheson, D. and B.M. Allos. 2001. Campylobacter jejuni infections: update on emerging issues and trends. Clinical Infectious Diseases. 32(8): 1201–6.

Allos, B.M. 2001. Campylobacter jejuni Infections: update on emerging issues and trends. Clinical infectious diseases : an official publication of the Infectious Diseases Society of America. 32(8): 1201–6.

Angulo, F.J., M.D. Kirk, I. McKay, G.V. Hall, C.B. Dalton, R. Stafford, L. Unicomb and J. Gregory. 2008. Foodborne disease in Australia: The OzFoodNet experience. Clinical Infectious Diseases. 47(3): 392–400.

Angulo, F.J., J.T. LeJeune and P.J. Rajala-Schultz. 2009. Unpasteurized milk: a continued public health threat. Clinical Infectious Diseases. 48(1): 93–100.

Ankolekar, C., T. Rahmati and R.G. Labbé. 2009. Detection of toxigenic *Bacillus cereus* and *Bacillus thuringiensis* spores in U.S. rice. International Journal of Food Microbiology. 128(3): 460–6.

Bae, J.Y., J.S. Lee, M.H. Shin, S.H. Lee and I.G. Hwang. 2011. Effect of wash treatments on reducing human norovirus on iceberg lettuce and perilla leaf. J. Food Prot. 74(11): 1908–11.

Baert, L., K. Mattison, F. Loisy-Hamon, J. Harlow, A. Martyres, B. Lebeau et al. 2011. Review: norovirus prevalence in Belgian, Canadian and French fresh produce: a threat to human health? Int. J. Food Microbiol. 151(3): 261–9.

Baker, R.C., M. Dulce, C. Paredes and R.A. Qureshi. 1987. Prevalence of Campylobacter jejuni in eggs and poultry meat in New York state. Poultry Science. 66(11): 1766–70.

Baker-Austin, C., L. Stockley, R. Rangdale and J. Martinez-Urtaza. 2010. Environmental occurrence and clinical impact of *Vibrio vulnificus* and *Vibrio parahaemolyticus*: a European perspective. Environmental Microbiology Reports. 2(1): 7–18.

Bennett, S.D., K.A. Walsh and L.H. Gould. 2013. Foodborne Disease Outbreaks Caused by *Bacillus cereus, Clostridium perfringens,* and *Staphylococcus aureus*—United States, 1998–2008. Clinical Infectious Diseases. 57(3): 425–33.

Besser, R.E., S.M. Lett, J.T. Weber, M.P. Doyle, T.J. Barrett, J.G. Wells and P.M. Griffin. 1993. An outbreak of diarrhea and hemolytic uremic syndrome from *Escherichia coli* O157:H7 in fresh-pressed apple cider. Jama. 269(17): 2217–20.

Bialek, S.R., P.A. George, G.-L. Xia, M.B. Glatzer, M.L. Motes, J.E. Veazey, R.M. Hammond, T. Jones, Y.C. Shieh and J. Wamnes. 2007. Use of molecular epidemiology to confirm a multistate outbreak of Hepatitis a caused by consumption of oysters. Clinical Infectious Diseases. 44(6): 838–40.

Black, R.E., R.J. Jackson, T. Tsai, M. Medvesky, M. Shayegani, J.C. Feeley, K.I. MacLeod and A.M. Wakelee. 1978. Epidemic Yersinia enterocolitica infection due to contaminated chocolate milk. New England Journal of Medicine. 298(2): 76–9.

Braden, C.R. and R.V. Tauxe. 2013. Emerging trends in foodborne diseases. Infectious Disease Clinics of North America. 27(3): 517–33.

Brynestad, S. and P.E. Granum. 2002. Clostridium perfringens and foodborne infections. International Journal of Food Microbiology. 74(3): 195–202.

Butt, A.A., K.E. Aldridge and C.V. Sanders. 2004. Infections related to the ingestion of seafood Part I: viral and bacterial infections. The Lancet Infectious Diseases. 4(4): 201–12.

Butzler, J.P. 2004. Campylobacter, from obscurity to celebrity. Clinical microbiology and infection: the official publication of the European Society of Clinical Microbiology and Infectious Diseases. 10(10): 868–76.

Chittick, P., A. Sulka, R.V. Tauxe and A.M. Fry. 2006. A summary of national reports of foodborne outbreaks of Salmonella Heidelberg infections in the United States: clues for disease prevention. Journal of Food Protection®. 69(5): 1150–3.

Collier, M.G., Y.E. Khudyakov, D. Selvage, M. Adams-Cameron, E. Epson, A. Cronquist et al. 2014. Outbreak of hepatitis A in the USA associated with frozen pomegranate arils imported from Turkey: an epidemiological case study. The Lancet Infectious Diseases. 14(10): 976–81.

Conaty, S., P. Bird, G. Bell, E. Kraa, G. Grohmann and J.M. McAnulty. 2000. Hepatitis A in New South Wales, Australia from consumption of oysters: the first reported outbreak. Epidemiology and Infection. 124(1): 121–30.

Control, CfD, Prevention. 2014. Foodborne outbreak online database. http://wwwncdcgov/foodborneoutbreaks/(accessed Jan 9, 2015).

Crim, S.M., M. Iwamoto, J.Y. Huang, P.M. Griffin, D. Gilliss, A.B. Cronquist, M. Cartter, M. Tobin-D'Angelo, D. Blythe and K. Smith. 2014. Incidence and trends of infection with pathogens transmitted commonly through food—foodborne diseases active surveillance network, 10 U.S. Sites, 2006–2013. MMWR: Morbidity & Mortality Weekly Report. 63(15): 328–32.

Cronquist, A.B., R.K. Mody, R. Atkinson, J. Besser, M.T. D'Angelo, S. Hurd, T. Robinson, C. Nicholson and B.E. Mahon. 2012. Impacts of culture-independent diagnostic practices on public health surveillance for bacterial enteric pathogens. Clinical Infectious Diseases. 54(suppl. 5): S432–S9.

Desenclos, J.C., K.C. Klontz, M.H. Wilder, O.V. Nainan, H.S. Margolis and R.A. Gunn. 1991. A multistate outbreak of hepatitis A caused by the consumption of raw oysters. American Journal of Public Health. 81(10): 1268–72.

Donado-Godoy, P., V. Clavijo, M. Leon, M.A. Tafur, S. Gonzales, M. Hume, W. Alali, I. Walls, D. Lo Fo Wong and M. Doyle. 2012. Prevalence of Salmonella on retail broiler chicken meat carcasses in Colombia. Journal of Food Protection®. 75(6): 1134–8.

Donado-Godoy, P., V. Clavijo, M. León, A. Arevalo, R. Castellanos, J. Bernal, M.A. Tafur, M.V. Ovalle, W.Q. Alali, M. Hume, J.J. Romero-Zuniga, I. Walls and M.P. Doyle. 2014. Counts, serovars, and antimicrobial resistance phenotypes of *Salmonella* on raw Chicken Meat at Retail in Colombia. Journal of Food Protection. 77(2): 227–35.

Donnan, E.J., J.E. Fielding, J.E. Gregory, K. Lalor, S. Rowe, P. Goldsmith, M. Antoniou, K.E. Fullerton, K. Knope and J.G. Copland. 2012. A Multistate Outbreak of Hepatitis A associated with semidried tomatoes in Australia, 2009. Clinical Infectious Diseases. 54(6): 775–81.

Doyle, M.P. and M.C. Erickson. 2008. Summer meeting 2007—the problems with fresh produce: an overview. Journal of Applied Microbiology. 105(2): 317–30.

Duhain, G.L., A. Minnaar and E.M. Buys. 2012. Effect of chlorine, blanching, freezing, and microwave heating on Cryptosporidium parvum viability inoculated on green peppers. Journal of Food Protection. 75(5): 936–41.

Escherichia coli O157:H7 infection associated with drinking raw milk—Washington and Oregon, November–December 2005, 2007. MMWR Morbidity and Mortality Weekly Report. 56(8): 165–7.

Escherichia coli O157:H7 infections in children associated with raw milk and raw colostrum from cows—California, 2006. 2008. MMWR Morbidity and Mortality Weekly Report. 57(23): 625–8.

European Food Safety Authority ECfDPaC. 2014. The European Union Summary Report on Trends and Sources of Zoonoses, Zoonotic Agents and Food-borne Outbreaks in 2012. EFSA Journal. 12(2): 3547.

Fankhauser, R.L., S.S. Monroe, J.S. Noel, C.D. Humphrey, J.S. Bresee, U.D. Parashar, T. Ando and R.I. Glass. 2002. Epidemiologic and molecular trends of "Norwalk-like viruses" associated with outbreaks of gastroenteritis in the United States. The Journal of Infectious Diseases. 186(1): 1–7.

Feldhusen, F. 2000. The role of seafood in bacterial foodborne diseases. Microbes and Infection. 2(13): 1651–60.

Gallot, C., L. Grout, A.-M. Roque-Afonso, E. Couturier, P. Carrillo-Santisteve, J. Pouey, M.-J. Letort, S. Hoppe, P. Capdepon and S. Saint-Martin. 2011. Hepatitis A associated with semidried tomatoes, France, 2010. Emerging Infectious Diseases 17(3): 566.

Gentry, J., J. Vinjé, D. Guadagnoli and E.K. Lipp. 2009. Norovirus distribution within an Estuarine Environment. Applied and Environmental Microbiology. 75(17): 5474–80.

Gillespie, I.A., G.K. Adak, S.J. O'brien and F.J. Bolton. 2003. Milkborne general outbreaks of infectious intestinal disease, England and Wales, 1992–2000. Epidemiology & Infection. 130(03): 461–8.

Global Foodborne Infections Network: WHO; 2014 [Nov 14]. Available from: http://www.who.int/gfn/goals/en/.

Gottlieb, S.L., E.C. Newbern, P.M. Griffin, L.M. Graves, R.M. Hoekstra, N.L. Baker, S.B. Hunter, K.G. Holt, F. Ramsey and M. Head. 2006. Multistate outbreak of listeriosis linked to turkey deli meat and subsequent changes in US regulatory policy. Clinical Infectious Diseases. 42(1): 29–36.

Gould, L.H., K.A. Walsh, A.R. Vieira, K. Herman, I.T. Williams, A.J. Hall and D. Cole. 2013. Surveillance for foodborne disease outbreaks—United States, 1998–2008. Morbidity and mortality weekly report Surveillance summaries (Washington, DC: 2002). 62(2): 1–34.

Graczyk, T.K., R.C. Thompson, R. Fayer, P. Adams, U.M. Morgan and E.J. Lewis. 1999. Giardia duodenalis cysts of genotype A recovered from clams in the Chesapeake Bay subestuary, Rhode River. The American Journal of Tropical Medicine and Hygiene. 61(4): 526–9.

Granum, P.E. and T. Lindbäck. 2013. *Bacillus cereus*. Food Microbiology: American Society of Microbiology.

Green, M. and G. Fitzsimmons. 2013. The OzFoodNet story: 2000 to present day. Microbiology Australia. 34(2): 59–62.

Greig, J.D. and A. Ravel. 2009. Analysis of foodborne outbreak data reported internationally for source attribution. International Journal of Food Microbiology. 130(2): 77–87.

Guerini, M.N., D.M. Brichta-Harhay, S.D. Shackelford, T.M. Arthur, J.M. Bosilevac, N. Kalchayanand, T.L. Wheeler and M. Koohmaraie. 2007. Listeria prevalence and *Listeria monocytogenes* serovar diversity at cull cow and bull processing plants in the United States. Journal of Food Protection®. 70(11): 2578–82.

Hall, R.L., A. Lindsay, C. Hammond, S.P. Montgomery, P.P. Wilkins, A.J. da Silva, I. McAuliffe, M. de Almeida, H. Bishop and B. Mathison. 2012. Outbreak of human trichinellosis in Northern california Caused by Trichinella murrelli. The American Journal of Tropical Medicine and Hygiene. 87(2): 297–302.

Hennessy, T.W., C.W. Hedberg, L. Slutsker, K.E. White, J.M. Besser-Wiek, M.E. Moen, J. Feldman, W.W. Coleman, L.M. Edmonson and K.L. MacDonald. 1996. A National Outbreak of Salmonella enteritidis infections from ice cream. New England Journal of Medicine. 334(20): 1281–6.

Herwaldt, B.L. and M.-L. Ackers. 1997. An outbreak in 1996 of cyclosporiasis cssociated with Imported Raspberries. New England Journal of Medicine. 336(22): 1548–56.

Hohweyer, J., A. Dumètre, D. Aubert, N. Azas and I. Villena. 2013. Tools and Methods for detecting and characterizing Giardia, Cryptosporidium, and Toxoplasma parasites in marine mollusks. Journal of Food Protection®. 76(9): 1649–57.

Holzbauer, S.M., W.A. Agger, R.L. Hall, G.M. Johnson, D. Schmitt, A. Garvey, H.S. Bishop, H. Rivera, M.E. de Almeida and D. Hill. 2014. Outbreak of Trichinella spiralis infections associated with a wild boar hunted at a game farm in Iowa. Clinical Infectious Diseases. 59(12): 1750–6.

Human tuberculosis caused by Mycobacterium bovis—New York City, 2001–2004. 2005. MMWR Morbidity and Mortality Weekly Report. 54(24): 605–8.

Hutin, Y.J., V. Pool, E.H. Cramer, O.V. Nainan, J. Weth, I.T. Williams, S.T. Goldstein, K.F. Gensheimer, B.P. Bell and C.N. Shapiro. 1999. A multistate, foodborne outbreak of hepatitis A. National Hepatitis A Investigation Team. The New England Journal of Medicine. 340(8): 595–602.

Incidence and trends of infection with pathogens transmitted commonly through food—foodborne diseases active surveillance network, 10 U.S. sites, 1996–2012. 2013 MMWR Morbidity and Mortality Weekly Report. 62(15): 283–7.

Iwamoto, M., T. Ayers, B.E. Mahon and D.L. Swerdlow. 2010. Epidemiology of Seafood-associated Infections in the United States. Clinical Microbiology Reviews. 23(2): 399–411.

Iwamoto, M., J.Y. Huang, A.B. Cronquist, C. Medus, S. Hurd, S. Zansky, J. Dunn, A.M. Woron, N. Oosmanally and P.M. Griffin. 2015. Bacterial Enteric Infections Detected by Culture-Independent Diagnostic Tests—FoodNet, United States, 2012–2014. MMWR Morbidity and Mortality Weekly Report. 64(9): 252–7.

Jaykus, L.A., D.H. Souza and C.L. Moe. 2013. Foodborne Viral Pathogens. Food Microbiology: American Society of Microbiology.

Johnson, E.A. 2013. Clostridium botulinum. Food Microbiology: American Society of Microbiology.

Kennedy, E.D., R.L. Hall, S.P. Montgomery, D.G. Pyburn and J.L. Jones. 2009. Trichinellosis surveillance—United States, 2002–2007. Morbidity and mortality weekly report Surveillance summaries (Washington, DC: 2002). 58(9): 1–7.

Kim, S.G., E.H. Kim, C.J. Lafferty and E. Dubovi. 2005. Coxiella burnetii in bulk tank milk samples, United States. Emerging Infectious Diseases. 11(4): 619–21.

Kuchenmüller, T., S. Hird, C. Stein, P. Kramarz, A. Nanda and A. Havelaar. 2009. Estimating the global burden of foodborne diseases—a collaborative effort. Euro surveillance: bulletin Européen sur les maladies transmissibles = European Communicable Disease Bulletin. 14(18).

Laboratory-confirmed non-O157 Shiga toxin-producing Escherichia coli—Connecticut, 2000–2005. 2007. MMWR Morbidity and Mortality Weekly Report. 56(2): 29–31.

Lanier, W.A., M.M. Leeper, K.E. Smith, G.E. Tillman, K.G. Holt and P. Gerner-Smidt. 2009. Pulsed-field gel electrophoresis subtypes of shiga toxin-producing Escherichia coli O157 isolated from ground beef and humans, United States, 2001–2006. Foodborne Pathogens and Disease. 6(9): 1075–82.

Lanier, W.A., J.M. Hall, R.K. Herlihy, R.T. Rolfs, J.M. Wagner, L.H. Smith and E.K. Hyytia-Trees. 2011. Outbreak of Shiga-toxigenic *Escherichia coli* O157:H7 infections associated with rodeo attendance, Utah and Idaho, 2009. Foodborne Pathogens and Disease. 8(10): 1131–3.

Laufer, A.S., J. Grass, K. Holt, J.M. Whichard, P.M. Griffin and L.H. Gould. 2014. Outbreaks of Salmonella infections attributed to beef—United States, 1973–2011. Epidemiology & Infection. FirstView: 1–11.

Li, H., H. Wang, apos, J.-Y. Aoust and J. Maurer. 2013. Salmonella Species. Food Microbiology: American Society of Microbiology.

Lianou, A. and J.N. Sofos. 2007. A review of the incidence and transmission of *Listeria monocytogenes* in ready-to-eat products in retail and food service environments. J. Food Prot. 70(9): 2172–98.

Lopez, A.S., D.R. Dodson, M.J. Arrowood, P.A. Orlandi, Jr., A.J. da Silva, J.W. Bier, S.D. Hanauer, R.L. Kuster, S. Oltman and M.S. Baldwin. 2001. Outbreak of cyclosporiasis associated with basil in Missouri in 1999. Clinical infectious diseases: an official publication of the Infectious Diseases Society of America. 32(7): 1010–7.

Luna-Gierke, R.E., P.M. Griffin, L.H. Gould, K. Herman, C.A. Bopp, N. Strockbine and R. Mody. 2014. Outbreaks of non-O157 Shiga toxin-producing *Escherichia coli* infection: USA. Epidemiology & Infection. 142(11): 2270–80.

Lynch, M.F., R.V. Tauxe and C.W. Hedberg. 2009. The growing burden of foodborne outbreaks due to contaminated fresh produce: risks and opportunities. Epidemiology & Infection. 137(Special Issue 03): 307–15.

MacKenzie, W.R., W.L. Schell, K.A. Blair, D.G. Addiss, D.E. Peterson, N.J. Hoxie, J.J. Kazmierczak and J.P. Davis. 1995. Massive outbreak of waterborne cryptosporidium infection in Milwaukee, Wisconsin: recurrence of illness and risk of secondary transmission. Clinical infectious diseases: an official publication of the Infectious Diseases Society of America. 21(1): 57–62.

Malek, M., E. Barzilay, A. Kramer, B. Camp, L.-A. Jaykus, B. Escudero-Abarca, G. Derrick, P. White, C. Gerba and C. Higgins. 2009. Outbreak of norovirus infection among river rafters associated with packaged delicatessen meat, Grand Canyon, 2005. Clinical Infectious Diseases. 48(1): 31–7.

Manning, S.D., R.T. Madera, W. Schneider, S.E. Dietrich, W. Khalife, W. Brown, T.S. Whittam, P. Somsel and J.T. Rudrik. 2007. Surveillance for Shiga toxin-producing *Escherichia coli*, Michigan, 2001–2005. Emerging Infectious Diseases. 13(2): 318–21.

McCollum, J.T., A.B. Cronquist, B.J. Silk, K.A. Jackson, K.A. O'Connor, S. Cosgrove, J.P. Gossack, S.S. Parachini, N.S. Jain and P. Ettestad. 2013. Multistate outbreak of listeriosis associated with cantaloupe. The New England Journal of Medicine. 369(10): 944–53.

McLaughlin, J.B., A. DePaola, C.A. Bopp, K.A. Martinek, N.P. Napolilli, C.G. Allison, S.L. Murray, E.C. Thompson, M.M. Bird and J.P. Middaugh. 2005. Outbreak of *Vibrio parahaemolyticus* gastroenteritis associated with Alaskan oysters. New England Journal of Medicine. 353(14): 1463–70.

Mead, P.S., E.F. Dunne, L. Graves, M. Wiedmann, M. Patrick, S. Hunter, E. Salehi, F. Mostashari, A. Craig and P. Mshar. 2006. Nationwide outbreak of listeriosis due to contaminated meat. Epidemiology & Infection. 134(04): 744–51.

Meng, J., J.T. LeJeune, T. Zhao and M.P. Doyle. 2013. Enterohemorrhagic *Escherichia coli*. Food Microbiology: American Society of Microbiology.

Moe, C.L. 2009. Preventing norovirus transmission: How should we handle food handlers? Clinical Infectious Diseases. 48(1): 38–40.

Nastasijevic, I., R. Mitrovic and S. Buncic. 2009. The occurrence of *Escherichia coli* O157 in/on faeces, carcasses and fresh meats from cattle. Meat Science. 82(1): 101–5.

Newton, A.E., N. Garrett, S.G. Stroika, J.L. Halpin, M. Turnsek and R.K. Mody. 2014. Notes from the field: Increase in Vibrio parahaemolyticus infections associated with Consumption of Atlantic Coast shellfish—2013. MMWR Morbidity and Mortality Weekly Report. 63(15): 335–6.

Notes from the field: use of electronic messaging and the news media to increase case finding during a Cyclospora outbreak—Iowa, 2013. MMWR Morbidity and Mortality Weekly Report. 62(30): 613–4.

Nygren, B.L., K.A. Schilling, E.M. Blanton, B.J. Silk, D.J. Cole and E.D. Mintz. 2013. Foodborne outbreaks of shigellosis in the USA, 1998–2008. Epidemiology and Infection. 141(2): 233–41.

Olsen, S.J., M. Patrick, S.B. Hunter, V. Reddy, L. Kornstein, W.R. MacKenzie, K. Lane, S. Bidol, G.A. Stoltman and D.M. Frye. 2005. Multistate outbreak of *Listeria monocytogenes* infection linked to delicatessen turkey meat. Clinical Infectious Diseases. 40(7): 962–7.

Ortega, Y.R. 2013. Protozoan Parasites. Food Microbiology: American Society of Microbiology.

Ortega, Y.R. and R. Sanchez. 2010. Update on *Cyclospora cayetanensis*, a Food-Borne and waterborne parasite. Clinical Microbiology Reviews. 23(1): 218–34.

Painter, J.A., T. Ayers, R. Woodruff, E. Blanton, N. Perez, R.M. Hoekstra, P.M. Griffin and C. Braden. 2009. Recipes for foodborne outbreaks: a scheme for categorizing and grouping implicated foods. Foodborne Pathogens and Disease. 6(10): 1259–64.

Painter, J.A., R.M. Hoekstra, T. Ayers, R.V. Tauxe, C.R. Braden, F.J. Angulo and P.M. Griffin. 2013. Attribution of foodborne illnesses, hospitalizations, and deaths to food commodities by using outbreak data, United States, 1998–2008. Emerging Infectious Diseases. 19(3): 407–15.

Park, S.Y., C.L. Woodward, L.F. Kubena, D.J. Nisbet, S.G. Birkhold and S.C. Ricke. 2008. Environmental dissemination of foodborne Salmonella in preharvest poultry production: Reservoirs, critical factors, and research strategies. Critical Reviews in Environmental Science and Technology. 38(2): 73–111.

Pires, S.M., A.R. Vieira, E. Perez, D.L.F. Wong and T. Hald. 2012. Attributing human foodborne illness to food sources and water in Latin America and the Caribbean using data from outbreak investigations. International Journal of Food Microbiology. 152(3): 129–38.

Potter, M.E., A.F. Kaufmann, P.A. Blake and R.A. Feldman. 1984. Unpasteurized milk: the hazards of a health fetish. Jama. 252(15): 2048–52.

Rangel, J.M., P.H. Sparling, C. Crowe, P.M. Griffin and D.L. Swerdlow. 2005. Epidemiology of *Escherichia coli* O157:H7 outbreaks, United States, 1982–2002. Emerging Infectious Diseases. 11(4): 603–9.

Ratnam, S., F. Stratton, C. O'Keefe, A. Roberts, R. Coates, M. Yetman, S. Squires, R. Khakhria and J. Hockin. 1999. Salmonella enteritidis outbreak due to contaminated cheese—Newfoundland. Canada communicable disease report = Releve des maladies transmissibles au Canada. 25(3): 17–9; discussion 9–21.

Rhoades, J.R., G. Duffy and K. Koutsoumanis. 2009. Prevalence and concentration of verocytotoxigenic *Escherichia coli*, Salmonella enterica and Listeria monocytogenes in the beef production chain: A review. Food Microbiology. 26(4): 357–76.

Rivera-Betancourt, M., S.D. Shackelford, T.M. Arthur, K.E. Westmoreland, G. Bellinger, M. Rossman, J.O. Reagan and M. Koohmaraie. 2004. Prevalence of *Escherichia coli* O157: H7, Listeria monocytogenes, and Salmonella in two geographically distant commercial beef processing plants in the United States. Journal of Food Protection®. 67(2): 295–302.

Robertson, L.J. and R.M. Chalmers. 2013. Foodborne cryptosporidiosis: is there really more in Nordic countries? Trends in Parasitology. 29(1): 3–9.

Robilotti, E., S. Deresinski and B.A. Pinsky. 2015. Norovirus. Clinical Microbiology Reviews. 28(1): 134–64.

Robins-Browne, R.M. 2013. Yersinia enterocolitica. Food Microbiology: American Society of Microbiology.

Ryan, C.A., M.K. Nickels, N.T. Hargrett-Bean, M.E. Potter, T. Endo, L. Mayer, C.W. Langkop, C. Gibson, R.C. McDonald and R.T. Kenney. 1987. Massive outbreak of antimicrobial-resistant salmonellosis traced to pasteurized milk. Jama. 258(22): 3269–74.

Ryser, E.T. and R.L. Buchanan. 2013. *Listeria monocytogenes*. Food Microbiology: American Society of Microbiology.

Salmonella typhimurium infection associated with raw milk and cheese consumption—Pennsylvania, 2007. 2007. MMWR Morbidity and Mortality Weekly Report. 56(44): 1161–4.

Salter, S.J. 2014. The food-borne identity. Nat. Rev. Micro. 12(8): 533-.

Samadpour, M., M.W. Barbour, T. Nguyen, T.M. Cao, F. Buck, G.A. Depavia, E. Mazengia, P. Yang, D. Alfi and M. Lopes. 2006. Incidence of enterohemorrhagic *Escherichia coli*, *Escherichia coli* O157, Salmonella, and *Listeria monocytogenes* in retail fresh ground beef, sprouts, and mushrooms. J. Food Prot. 69(2): 441–3.

Scallan, E., P.M. Griffin, F.J. Angulo, R.V. Tauxe and R.M. Hoekstra. 2011. Foodborne illness acquired in the United States—unspecified agents. Emerging Infectious Diseases. 17(1): 16.

Scallan, E., R.M. Hoekstra, F.J. Angulo, R.V. Tauxe, M.A. Widdowson, S.L. Roy et al. 2011. Foodborne illness acquired in the United States—major pathogens. Emerging Infectious Diseases. 17(1): 7–15.

Shayegani, M., D. Morse, I. DeForge, T. Root, L.M. Parsons and P. Maupin. 1983. Microbiology of a major foodborne outbreak of gastroenteritis caused by Yersinia enterocolitica serogroup O: 8. Journal of Clinical Microbiology. 17(1): 35–40.

Smith, S., J. Maurer, A. Orta-Ramirez, E. Ryser and D. Smith. 2001. Thermal inactivation of *Salmonella* spp., Salmonella typhimurium DT104, and *Escherichia coli* O157: H7 in ground beef. Journal of Food Science. 66(8): 1164–8.

Sofos, J.N. 2008. Challenges to meat safety in the 21st century. Meat Science. 78(1): 3–13.

Sofos, J.N., G. Flick, G.-J. Nychas, apos, C.A. Bryan, S.C. Ricke et al. 2013. Meat, Poultry, and Seafood. Food Microbiology: American Society of Microbiology.

Soler, P., S. Herrera, J. Rodríguez, J. Cascante, R. Cabral, A. Echeita-Sarriondia et al. 2008. Nationwide outbreak of Salmonella enterica serotype Kedougou infection in infants linked to infant formula milk, Spain, 2008. Euro surveillance: bulletin Europeen sur les maladies transmissibles= European Communicable Disease Bulletin. 13(35): 717–27.

Surveillance for foodborne disease outbreaks—United States, 2009–2010. 2013. MMWR Morbidity and Mortality Weekly Report. 62(3): 41–7.

Tacket, C.O., J.P. Narain, R. Sattin, J.P. Lofgren, C. Konigsberg, R.C. Rendtorff et al. 1984. A multistate outbreak of infections caused by Yersinia enterocolitica transmitted by pasteurized milk. Jama. 251(4): 483–6.

The European Union summary report on trends and sources of zoonoses, zoonotic agents and food-borne outbreaks in 2010, 2012. Euro surveillance : bulletin Europeen sur les maladies transmissibles = European Communicable Disease Bulletin. 17(10).

Threlfall, E.J., J. Wain, T. Peters, C. Lane, E. De Pinna, C.L. Little et al. 2014. Egg-borne infections of humans with salmonella: not only an *S. enteritidis* problem. World's Poultry Science Journal. 70(01): 15–26.

Tian, P., D. Yang and R. Mandrell. 2011. Differences in the binding of human norovirus to and from romaine lettuce and raspberries by water and electrolyzed waters. J. Food Prot. 74(8): 1364–9.

Two multistate outbreaks of Shiga toxin-producing *Escherichia coli* infections linked to beef from a single slaughter facility—United States, 2008, 2010. MMWR Morbidity and Mortality Weekly Report. 59(18): 557–60.

Uyttendaele, M., S. Vankeirsbilck and J. Debevere. 2001. Recovery of heat-stressed *E. coli* O157:H7 from ground beef and survival of *E. coli* O157:H7 in refrigerated and frozen ground beef and in fermented sausage kept at 7°C and 22°C. Food Microbiology. 18(5): 511–9.

Wallace, B.J., J.J. Guzewich, M. Cambridge, S. Altekruse and D.L. Morse. 1999. Seafood-associated disease outbreaks in New York, 1980–1994. American Journal of Preventive Medicine. 17(1): 48–54.

Warren, B.R., H.-G. Yuk and K.R. Schneider. 2007. Survival of Shigella sonnei on smooth tomato surfaces, in potato salad and in raw ground beef. International Journal of Food Microbiology. 116(3): 400–4.

Wheeler, C., T.M. Vogt, G.L. Armstrong, G. Vaughan, A. Weltman, O.V. Nainan et al. 2005. An outbreak of hepatitis A associated with green onions. The New England Journal of Medicine. 353(9): 890–7.

Wilson, N.O., R.L. Hall, S.P. Montgomery and J.L. Jones. 2015. Trichinellosis Surveillance—United States, 2008–2012. Morbidity and mortality Weekly Report Surveillance summaries (Washington, DC: 2002). 64(Ss-01): 1–8.

Zhao, C., B. Ge, J. De Villena, R. Sudler, E. Yeh, S. Zhao et al. 2001. Prevalence of *Campylobacter* spp., *Escherichia coli*, and Salmonella Serovars in Retail Chicken, Turkey, Pork, and Beef from the Greater Washington, D.C., Area. Applied and Environmental Microbiology. 67(12): 5431–6.

CHAPTER 6

Molecular Methodologies for Investigating Foodborne Infection

Sarah A. Waldman[1] and *Stuart H. Cohen*[2,*]

Introduction

The rapid identification of and "fingerprinting" bacterial pathogens is an essential component of epidemiological surveillance and investigating the outbreak of a foodborne infection in order to improve food safety. Foodborne outbreaks are increasingly common. In the last two years the CDC has performed at least 25 multistate foodborne outbreak investigations (http://www.cdc.gov/foodsafety/outbreaks/multistate-outbreaks/ outbreaks-list.html). During the last two decades, molecular methods of bacterial strain identification have replaced phenotypic assays, increasing the discriminatory ability of identifying and typing the pathogenic bacterial strains. They have transformed the way an outbreak investigation is conducted. Standard bacterial phenotyping uses the morphology of bacterial colonies on culture media, as well as serological and biochemical assays, and antibiotic susceptibilities, to distinguish pathogenic microorganisms and identify different strains (Arbeit 1995). However, these traditional methods lack the ability to distinguish between two closely related bacterial strains sensitively. Further limitations of this approach include poor reproducibility,

---

[1] Fellow in Infectious Diseases, Division of Infectious Diseases, University of California, Davis, Sacramento, CA 95817.
[2] Professor and Chief, Division of Infectious Diseases, University of California, Davis, Sacramento, CA 95817.
* Corresponding author: stcohen@ucdavis.edu

need for specialized reagents and the lack of applicability to a wide range of bacterial pathogens. Molecular analysis differentiates between bacterial strains on the basis of their genetic profile. However, these approaches can be costly and time-intensive and require standardization (Field et al. 2004).

The analysis of proteins and nucleic acids have enabled enhanced characterization of bacterial species, a greater understanding of bacterial diversity and improved laboratory-based surveillance for the emergence of new strains (Hahm et al. 2003, Lim et al. 2005). The superior timeliness and specificity of laboratory data provided by the molecular identification techniques have changed the depth of understanding fundamentally and transformed the epidemiology of outbreak investigation. However, the large volume of data has presented new challenges including increased requirements for international standardization among different laboratories, new data analysis and storage needs and investment in workforce development. This chapter will discuss the basic principles of some of the most commonly used molecular methods for the typing of foodborne bacterial infection including some of the newer methodologies. We will review their applicability in outbreak investigation.

Molecular Strain Typing

Molecular strain typing refers to methods developed to identify bacterial strains using analysis of DNA. These techniques examine specific characteristics of a group of organisms that provide the ability to differentiate between them beyond the species or serotype level. In outbreak investigations, molecular data is used to assess the genetic relatedness between suspected microbes in order to establish which ones are indistinguishable from one another and may be part of the outbreak. This classification helps investigators focus on an investigation saving both time and resources.

There are several ways that DNA analysis can be used for this aim and can generally be classified into two categories: (1) DNA banding patterns, (2) DNA sequencing-methods (Li et al. 2009). DNA banding pattern-based genotyping methods differentiate strains based on the size of the DNA fragments formed by cutting DNA using restriction enzymes (REs). DNA sequencing-based methods discriminate polymorphisms in the bacterial strain DNA using the consensus sequence of specific genes or of the whole organism. DNA hybridization-based methods are a subset of DNA sequencing methods that identify bacterial strains based on the binding of their DNA to the sequences of known probes, commonly known as DNA arrays. The novel methodology of next generation sequencing (NGS), allowing millions of sequencing reactions to run parallel, can identify single-nucleotide variants (SNVs) and has been used to type organisms like *Clostridium difficile* (Eyre et al. 2013). This method also has applications in foodborne epidemic surveillance as will be discussed later in the chapter.

As technology advances at an accelerating rate, the collection of molecular methodologies is expanding, increasing the discriminatory capability, reproducibility and repeatability of bacterial strain identification. There are several practical concerns about how these assays are implemented and used that need to be considered, including speed, expense, data output at the end of the analysis, strength, and generalizability, ease of use of the assay, and the simplicity of data analysis and sharing. It will ultimately be useful and even necessary to develop international practices to provide standardization between labs.

Methods of Bacterial Strain Typing

DNA Fragment Based Analysis

In this approach, DNA fragments are produced either using restriction enzymes to cleave DNA, or PCR is used to amplify specific DNA regions, or a combination of both methods. Enzymatic restriction identifies and cuts target DNA precisely at a defined sequence location. Alternatively DNA amplification using specific primers can create millions of copies of a specific fragment, making it a sensitive, rapid and generally applicable tool for human and environmental specimens (Lo and Chan 2006).

Restriction Fragment Pulsed-Field Gel Electrophoresis (PFGE) Analysis

Described first in 1984, PFGE is currently among the most frequently used DNA-based typing methods in the setting of foodborne bacterial outbreak identification, investigation and surveillance. This methodology has been used to identify several outbreaks of the common foodborne pathogens *Salmonella enterica* strains and the Shiga-toxin-producing *E. coli* (STEC) (Schwartz et al. 1984, Ribot et al. 2006). This method separates large DNA molecules (10kb–10Mb) to generate identifiable patterns. In general, the bacterial isolates are prepared for electrophoresis after being cultured. The organisms are suspended in agarose and subsequently lysed with specific enzymes and detergents to prevent degradation of the genomic DNA during the extraction process. Restriction endonucleases that cut the DNA infrequently are then used to digest the samples, and DNA is separated on agarose gel. By using a rotating electrical field gradient parallel with the DNA movement, large fragments of DNA, which account for the majority of the genome, can progress through the agarose gel matrix to provide a fingerprint of the organism. Although DNA banding patterns are associated with different bacterial subtypes, the specific DNA content of the bands is unknown. The choice of the restriction endonuclease is crucial in determining reliable PFGE banding patterns, as each restriction enzyme target site is unique. The more complex banding patterns increase the discriminatory power of PFGE (Chen et al. 2005). The resulting band pattern is analyzed with specialized image analysis software and compared against standards, or *in silico* (computer-based) searches of completed bacterial sequences to identify specific patterns.

Using the approach of PFGE for bacterial strain identification has a high discriminatory capability for most pathogens considered, including those in which a complete understanding of the molecular biology of the bacteria may be absent. Set protocols have been developed and rigid quality control introduced for many high-priority pathogens with the goal of developing international standards among laboratories and increasing reproducibility. PulseNet International, a global PFGE database that collects DNA sequences of different foodborne bacterial strains in order to assist with outbreak investigation, has contributed a large part to the quality control of PFGE techniques as well as popularity of PFGE use for specific bacterial pathogens. Data acquired from PFGE techniques and subsequently sent to PulseNet for analysis by microbiologists and epidemiologists were essential for the diagnosis of outbreaks of *Salmonella enterica* Typhimurium and Saintpaul infections associated with unpasteurized orange juice (Jain et al. 2009), *Listeria monocytogenes* associated with cantaloupe (McCollum et al. 2013), and STEC infections associated with beef (CDCP 2010).

Although the reproducibility and repeatability of data acquired are strengths of PFGE, extensive time involved and labor intensity are weaknesses. Additionally, there are some strains that cannot be typed using PFGE techniques, including *Salmonella enterica* Typhimurium DT 104, and *Salmonella enterica* Enteritidis PT 4. PFGE lacks the resolution power to distinguish between bands of similar size (Davis et al. 2003). It also requires high-quality DNA. Other shortcomings include the risk of laboratory contamination prior to applying the proteases and restriction enzymes, as well as many other factors that may influence band pattern, such as difference in agarose gel composition and temperature, and applied electrical voltage (Chung et al. 2000).

Restriction Fragment Length Polymorphism (RFLP) Analysis

RFLP uses a similar approach to PFGE, but targets small fragments of known sequences using fluorescent (or radioactive) probes to target specific sequences. RFLP measures the size of restriction fragments that generated target specific DNA sequences. It was one of the first techniques used to measure variation in DNA sequence (Thibodeau 1987). Rather than using restriction endonucleases that cut rarely, RFLP employs frequently cutting restriction endonucleases to digest genomic DNA resulting in hundreds of short restriction fragments that are then separated by gel electrophoresis and hybridized with probes for visualization of specific banding patterns that are later analyzed and compared to a bacterial strain databases. This method can be used after bacterial isolates have been cultured from human or environmental samples. Alternatively, total DNA can be extracted from the specimen followed by locus specific PCR amplification and subsequently subjected to RFLP. Differences in the pattern of restriction fragments reflect the variable distances between cleavage sites of bacterial strains, allowing for identification of genetic similarity (Busse et al. 1996). One of the limitations of RFLP is the requirement of high quality genomic DNA, which makes its application in many cases of outbreak investigation difficult.

Ribotyping

Ribotyping employs similar techniques to RFLP to evaluate polymorphisms in bacterial genes coding for 16S and 23S rRNA (Bingen et al. 1994). Bacterial rRNA operons are a group of highly conserved genes and are flanked by more variable gene sequences. Variation in gene sequence of the flanking restriction sites lead to differing banding patterns (ribotypes) after restriction digestion when probed with the conserved portions of 16S and 23S rRNA genes. Ribotyping allows for analysis without prior knowledge of unique genomic sequence for each subtype. It has been utilized as an epidemiologic tool for typing of *E. coli* Serogroup O157 isolates as well as subtyping *Salmonella enterica* Enteritidis strains among other bacteria (Clark et al. 2003, Martin et al. 1996).

Locus Specific Amplification Analysis

Bacterial genomes contain multiple short nucleotide sequences that are organized with a variable number of tandem (end to end) repeat (VNTR) sequences. These loci are dispersed throughout bacterial genomes, in both coding and non-coding regions (Lindstedt 2005, Lupski and Weinstock 2000, van Gent et al. 2011). Multiple-Locus Variable tandem repeat Analysis

(MLVA) assays measure the length of variable tandem repeat sequences by locus-specific PCR amplification followed by separation by capillary gel electrophoresis. The patterns of locus-specific repeat sequence lengths are used to create a specific genotype or fingerprints that discriminate between isolates of the same species (Keim et al. 2000). Multi-locus variable number tandem repeat analysis has been used in the identification of multiple foodborne pathogen outbreaks as an alternative to PFGE, including *E. coli* O157:H7 (Noller et al. 2003, Hyytia-Trees et al. 2006, Byrne et al. 2014), *E. coli* O26 strains (Miko et al. 2010), *S. enterica* Typhimurium (Larsson et al. 2009), and *S. enterica* Enteritidis (Bertrand et al. 2015).

There are multiple advantages to MLVA analyses. It has the higher level of discrimination necessary to distinguish between various strains of bacteria when compared to other typing methods (Lindstedt 2005, Tenover et al. 2007). MLVA is able to distinguish between strains of Shiga toxin-producing *E. coli*, *Salmonella enterica*, and *L. monocytogenes*, with high resolution, although the variability among *Campylobacter jejuni* subtypes is not adequate using this approach. The simple procedure for MLVA and the ability to translate the results into a predefined numeric string of data are the advantages of this method. For *S. enterica* Typhimurium, Larsson and colleagues showed that results of testing strains in a reference collection using set guidelines, are highly reproducible and permit accurate inter-laboratory comparisons. They note that it is essential to take into account differences among laboratories. Calibration of fragment size must be included during analysis in each laboratory (Larsson et al. 2009). In another study, Larsson and colleagues described a tool for laboratories to compare their MVLA results regardless of the hardware, primers or settings in which they are employed. By using a set of calibration strains to evaluate fifteen typed strains of *S. enterica* Typhimurium with MLVA methodology at 20 international participating laboratories, inter-laboratory results were comparable despite using different conditions and hardware (Larsson et al. 2013). MVLA can produce rapid results, has high discriminatory power and the potential for international comparability between laboratories. Thus it is one of the more useful strategies for differentiating between bacterial strains. Conversely, the variable number of tandem repeat loci may evolve too rapidly to identify reliable relationships between related strains over a period of time.

DNA Sequencing

The methods described above have played an essential role in establishing approaches to the molecular epidemiology of foodborne pathogen. Extensive experience with these approaches has created large data sets for comparison when a new outbreak investigation is initiated. However, DNA sequencing technologies are quickly becoming the gold standard for identifying differences in bacterial strains in order to track, evaluate and contain foodborne disease outbreaks. The procedures and platforms for DNA sequencing have evolved quickly to allow more accurate, more sensitive and especially more economically feasible approaches to strain typing. The nucleotide sequence of a genomic fragment will directly determine identification of sub-strains and when compared to historical references and published databases, demonstrate the ability to differentiate and cluster isolates. Sequence-based methods permit high interlaboratory reproducibility, since sequence data can be easily compared among laboratories. Centrally managed genomic databases can collect large DNA locus-specific sequences from bacteria allowing for ease of exchange and development of consensus typing standards. From these storage collections, molecular microbiologists can subsequently use well-established analytical methods to assist with bacterial strain comparison in the setting of an outbreak investigation without detailed knowledge of the standard procedures being used in distributed analytic laboratories.

Until recently, Sanger sequencing (Sanger et al. 1977) has been the standard approach for bacterial sequencing. Many of the next generation methodologies must still validate results by comparing them with the Sanger method. Sanger DNA sequencing uses dideoxynucleotides that are randomly inserted into targeted DNA fragments during DNA synthesis, halting chain elongation at each dideoxynucleotide. A set of polynucleotide fragments of varying lengths are created, allowing the genomic sequence to be inferred from the length of the fragment and type of dideoxynuclotide at the terminus. This method is robust, easy to interpret with clear standardization, and multiple platforms exist for clinical laboratory use. However, significant amounts of DNA are necessary for analysis and the sensitivity is limited to ~ 20% in a total DNA mixture. Newer methods of DNA sequencing, described below, are more sensitive, but require more complicated procedures, costlier reagents and automated platforms.

Single locus sequence typing (SLST) methods use a Sanger sequencing approach to evaluate a single gene or genetic locus that is known to carry sufficient polymorphism to allow differentiation within a species and be useful as a typing approach. The single locus examined can be useful in comparing bacterial strains and determining phylogenetic relationships. More commonly, however, multiple loci can be examined, referred to as multi locus sequence typing (MLST), using a similar methodology (see below). Sequencing of the flaA gene has been used to differentiate isolates of *Campylobacter jejuni* (Meinersmann et al. 1997). In other bacterial populations, sequencing of the spa gene (an 8-codon variable number tandem repeat at the end of the gene for staphylococcal protein A) is one of the approaches for strain typing of methicillin-resistant *Staphylococcus aureus* (Frénay et al. 1994) particularly when paired with additional confirmatory assays. Scheutz and colleagues compared Shiga toxin (Stx) gene sequences in shiga-toxing producing *E. coli* strains for phylogenetic sequence relatedness resulting in a protocol to identify the subtypes Stx1 and Stx2 using PCR and a description of a standardized nomenclature (Scheutz et al. 2012). SLST has the potential for high discriminatory power, particularly for Shiga toxin-producing *E. coli* (STEC) and *Campylobacter* species, however it's capabilities are species dependent. The data from SLST are reproducible and reliable and typing results can be available within 24 hours.

Multi locus sequence typing (MLST) describes the sequencing of multiple genetic loci in several conserved genes (usually seven) and is a commonly used genotyping method for bacterial strain identification (Maiden et al. 1998). The genetic loci are usually part of genes that encode for "housekeeping genes", or essential proteins engaged in metabolic function of the organism and are typically conserved in all isolates. Fragments of 450–500 bp of seven genes are sequenced and each sequence for the given gene is assigned a number based on the pattern of single nucleotide polymorphisms (SNPs) detected. Each strain is then given a seven number profile representing the allelic profile or sequence type (Maiden et al. 1998). The sequence types are then assigned consolidated numbers so that the allelic profile can be summarized in a single number. MLST is useful for subtyping bacterial population with significant genetic recombination, such as *Neisseria meningitides* (Maiden et al. 1998), *Enterococcus faecalis* (Ruiz-Garbajosa et al. 2006). It has also been used for several foodborne pathogens, such as *C. jejuni* (Korczak et al. 2009) *E. coli* (Wirth et al. 2006), and *L. monocytogenes* (Revazaishvili et al. 2004).

MLST DNA sequences are stored in online databases making the information easily available for strain typing and data analysis among laboratories. Additional information is available at a standard MLST webpage, which currently stores MLST allele sequence data for 21 bacteria (http://www.mlst.net). One of the advantages of this technique is that the data can be analyzed in many different ways to examine the structure within MLST data collections and establish relationships between strain types. MLST methodology can also be used to evaluate virulence genes rather than housekeeping genes and this method called Multi-virulence-locus sequence typing (MVLST) can be used to supplement MLST for increased resolution. One of the drawbacks of MLST sequencing is that the DNA profile of the conserved housekeeping genes may not reveal adequate differences between similarly related bacterial isolates, such as with *S. enterica* Typhi, as well as being a costly and time intensive method.

Multispacer typing (MST) uses the highly variable non-coding DNA sequences of bacteria as markers for strain typing. It is a method of sequencing variable areas of the genome by using non-variable sequences to amplify the DNA. After selection of highly variable regions of interest, amplification is performed using primers in flanking genes; the PCR product is then sequenced and subsequently analyzed. First described with *Yersinia pestis* in 2004, this method takes advantage of the observation that the intergenic sequences of DNA have less selection pressure than the genetic material responsible for proteins (Drancourt et al. 2004). Comparison of MST to MLST has demonstrated that the former has been noted to have higher resolution (Fournier et al. 2004, Li et al. 2006). As the increasing number of bacterial species have more than one genome sequence available, the highly variable intergenic components are becoming more powerful for bacterial typing and MST genotypes are available for analysis through online databases.

DNA hybridization-based genotyping techniques apply probes and targets to detect DNA mutations using sequence complementarity. DNA targets, which are the free nucleic acids, are detected by hybridization of fluorescently labeled probes, DNA fragments of varying length and known sequence, by identification of complementary sequences. This methodology is performed on a substrate array, usually glass, plastic or metal alloys, allowing for genomic DNA to be tested for its ability to hybridize to thousands of known DNA sequences. Each spot on an array contains multiple identical strains of DNA with a unique sequence and each spot represents one gene. There are thousands of spots arrayed in orderly rows and columns on the substrate. There are two classes of DNA arrays, macroarrays and microarrays, which differ in the size and number of testing spots available. Macroarrays contain up to 5000 spots making it a less expensive technique whereas microarrays are denser with up to 10^6 spots per array with a higher discriminatory power. DNA hybridization based technology has the potential for high-throughput detection of multiple pathogens in one experiment. DNA microarray methodology has been applied for specific detection of foodborne pathogens after multiplex PCR (Bodrossy and Sessitsch 2004), and using oligonucleotide microarray methods (Sergeev et al. 2004, Eom et al. 2007, Kim et al. 2008).

Whole Genome Sequencing

Since the introduction of next-generation sequencing techniques over the last decade, most of the prominent novel methodologies focus on rapid sequencing of whole genomes using different principles. The major hardware platforms currently in use: (1) Life Technologies IonTorrent, which produces shorter sequences, (2) Roche 454, which employs pyrosequencing to generate longer read lengths but in smaller numbers, (3) Illumina sequencing technology, which produces shorter sequences with high throughput, (4) Pacific Biosciences SMRT sequencing system, which produces long sequences with higher error rates (EFSA 2013). The performance characteristics of the underlying technology vary, depending on read lengths, sequence outputs, error rates and time used often by several orders of magnitude (Chan et al. 2012, Loman et al. 2012). The end result of these technologies is the entire genome sequencing of the pathogen, providing the maximum approach to differentiate bacterial isolates.

One subset of whole genome sequencing for bacterial typing involves variant detection of SNPs between isolates. Initially unaltered whole-genome sequence data from the isolates in question are evaluated, cropped and mapped against a standard reference genome (usually a closely related, completed genome from a public database). However, *de novo* assemblies of entire genomes can be employed from epidemiologically identified isolates (e.g., from the putative index case in an outbreak investigation) (Price et al. 2012). After mapping the organism, variance in SNPs can be detected relative to the reference genome. The distribution of SNPs across the genome can be categorized in each variant position (in reference to gene versus non-gene component). Screening is performed both on the individual and population level in order to identify different informative SNPs that assist with strain identification. This method involves significant data production that needs to be interpreted, analyzed, and stored (van Gent et al. 2011). One limitation for the use of genome SNP typing is that in species with genomic diversity, such as *Neisseria*, mapping against a historical reference strain may not be relevant for present-day isolates because of genomic divergence.

More recent DNA sequencing technologies are examining detection of single-nucleotide differences in a single cell. This technology is being incorporated into commercially available hardware by Oxford Nanopore technology.

Next-generation sequencing (NGS), which is commonly used to refer to whole genome sequencing methods, has been successfully employed on foodborne bacterial pathogens highlighting their potential for use in active outbreak investigation. Lambert and colleagues have published a proof-of-concept study describing the use of a whole genome sequencing approach that produced same-day results allowing in the identification of specific virulence marker genes of STEC (Lambert et al. 2015). Whole genome sequencing has helped in the understanding of mechanisms contributing to increased virulence among outbreaks of foodborne bacterial pathogens, including STEC, *S. enterica* subtype Montevideo

(Lienau et al. 2011), *S. enterica* Enteritidis (Deng et al. 2015), *S. enterica* Typhimirium DT 8 (Ashton et al. 2015), and *C. botulinum* (Mazuet et al. 2015).

Recent technological advances have reduced the cost of using whole genome sequencing technology in foodborne outbreak investigation. Molecular methods, such as next-generation sequencing, do not require prior culture and have improved ability to identify nonviable or fastidious organisms from direct specimens. However, in a polymicrobial specimen, typically only the predominant organism will be identified. One of the major challenges for the use of whole genome sequencing in outbreak investigation lies in the massive volume of data created that must be analyzed, assembled, interpreted and compared. Approaches have been developed to use whole genome sequencing to create publicly available large genetic databases of foodborne pathogens (e.g., 100 K Foodborne pathogen genome project[1]). Given the rapid development of whole genome sequencing, this technology is likely to be used in reference laboratories in the near foreseeable future, although it is unclear which technologies will be used regularly and how the data will be used.

PulseNet

In the current setting where identification and strain typing of pathogenic bacteria produces massive amounts of complex data, a central genomic database is an essential component of modern public health disease surveillance and important to outbreak response. PulseNet, the national network for comparing molecular strains of common foodborne bacterial pathogens run by the Centers for Disease Control and Prevention, has been invaluable in defining outbreaks in the United States. The network, which began in 1996 typing a single pathogen of *E. coli* O157:H7 with 10 laboratories involved, now includes all 50 states and 87 federal, regional state and local laboratories (Swaminathan et al. 2001). In providing testing methods, technology and data needed to identify foodborne outbreaks, PulseNet has standardized molecular strain identification of many common foodborne pathogenic organisms, including *E. coli* O157 and other STEC, *Campylobacter jejuni*, *Clostridium botulinum*, *Listeria monocytoe genes*, *Salmonella enterica* serotypes, *Shigella* species, *Vibrio cholera*, and *Vibrio parahaemolyticus*. PFGE and MLVA are the main strain typing methods used by PulseNet and the network has standard protocols for each bacteria.

PulseNet has been influential in the epidemiologic surveillance and identification of many foodborne outbreaks. After a patient presents with

---

[1] Further information on the 100 K foodbrone pathogen genome project available at: http://100kgenome.vetmed.ucdavis.edu/ (last visited on 4/12/2015).

diarrhea, the physician orders a diagnostic test (usually a stool culture) to determine the etiologic agent. The bacterial isolate's unique fingerprint, standardized to shared reference controls, is subsequently sent to an electronic database where microbiologists and epidemiologists can review the laboratory results to identify any strains that might suggest a cluster outbreak warranting further investigation. By providing essential standards and procedures, PulseNet ensures the quality and uniformity of the data that can be employed to identify cases that are part of an outbreak or detecting outbreaks through surveillance.

Recently PulseNet has employed newer strategies, such as culture-independent diagnostic practices. For decades laboratories have performed diagnostic testing for pathogens such as *Salmonella*, *Campylobacter* and STEC from culture. The paradigm is changing as laboratories adopt non-culture based methods for antigens, nucleic acid or toxins as they face the challenges of providing reliable and rapid test results with lower costs (Cronquist et al. 2012). Shiga toxin testing for the diagnosis of STEC has been available for decades (Hoefer et al. 2011). Rapid and validated PCR tests are available for *Campylobacter jejuni*, *Salmonella enterica*, *Yersinia* species and enteroinvasive *E. coli* from stool samples (Cunningham et al. 2010). There is an FDA-approved enzyme immunoassay to identify *Campylobacter jejuni* infection. Culture-independent testing has often improved sensitivity for some pathogens. Due to rapid detection there is increased potential for use, which can lead to identification of further cases (Cronquist et al. 2012). These methods provide different challenges. They typically do not provide antimicrobial susceptibility that can lead to difficulties in treating some patients. The increased sensitivity can identify colonized patients that do not have infection with the potential for increased investigation of cases causing wasted resources and unnecessary follow-up. This can also lead to overestimates of actual number of illnesses. Finally, the lack of an organism may lead to the loss of the ability to perform strain typing important for outbreak detection and control.

Whereas the "gold standard" PFGE provides results in 3–5 days, NGS can provide results within 1–2 days although interpretation of the ensuing data may take hours to days and automated informatics software must still be developed and validated for this application. PFGE has been well established to differentiate and cluster isolates. While the potential for NGS is great, it needs to be proven. Collaborations within the CDC, FDA and UC Davis in the 100 K Foodborne Pathogens Project are using NGS sequence strains from historical and recent outbreaks as well as strains from non-outbreaks to determine utility. Next generation sequencing has provided PulseNet with a potential tool for surveillance and outbreak investigation that can address some of the challenges noted above. The automation and potential for genetic serotyping, virulence, antimicrobial resistance profiling and strain typing make NGS an important novel technology to incorporate into PulseNet's procedures.

Conclusion

This chapter has focused on most of the widely used methods for molecular typing of foodborne pathogens. Currently, molecular typing of foodborne pathogens is performed with one or more of the methods from the wide array of techniques, PFGE in particular. The widespread availability together with simple bioinformatic analysis of the data acquired makes these methods attractive approaches for bacterial strain typing despite some of their limitations. These methods have generated large databases of information that can be accessed.

Whole genome sequencing by NGS appears to be the future direction of bacterial pathogen typing. Whole genome sequencing with the addition of culture-independent techniques has the potential for automation and genetic serotyping, virulence, antimicrobial resistance profiling and strain typing, making these technologies important tools in identification for surveillance and outbreak investigation of foodborne pathogens. In addition to the standardization of current technology, the bioinformatics and workforce structure to modify, interpret, and analyze large volumes of data need to be developed.

In short, the future of molecular typing methods is in providing high discriminatory power, reliable and reproducible data, that is easy to collect, share, and interpret. This provides the best opportunity to define the epidemiology of these infections and provide the necessary information to intervene on the transmission, preventing future infections. To date, none of the available methods fulfill all of these requirements and therefore a combination of the approaches must be used to address the identification, control and prevention of outbreaks optimally.

References

Arbeit, R.D. 1995. Laboratory procedures for the epidemiologic analysis of microorganisms. pp. 116–137. *In*: P.R. Murray, E.J. Baron, M.A. Pfaller, F.C. Tenover and R.H. Yolken (eds.). Manual of Clinical Microbiology. Washington DC: ASM Press.

Ashton, P.M., T. Peters, L. Ameh, R. McAleer, S. Petrie, S. Nair, I. Muscat, E. de Pinna and T. Dallman. 2015. Whole Genome Sequencing for the Retrospective Investigation of an Outbreak of Salmonella Typhimurium Dt 8. PLoS Curr. 7.

Bertrand, S., G. De Lamine de Bex, C. Wildemauwe, O. Lunguya, M.F. Phoba, B. Ley, J. Jacobs, R. Vanhoof and W. Mattheus. 2015. Multi Locus Variable-Number Tandem Repeat (Mlva) Typing Tools Improved the Surveillance of Salmonella Enteritidis: A 6 Years Retrospective Study. PLoS One 10, no. 2: e0117950.

Bingen, E.H., E. Denamur and J. Elion. 1994. Use of ribotyping in epidemiological surveillance of nosocomial outbreaks. Clin. Microbiol. Rev. 7(3): 311–27.

Bioscience, Meridian. Meridian Bioscience Receives Fda Clearance for New Campylobacter Test. http://www.meridianbioscience.com/Content/Assets/Files/2.2%20Foodborne%20Products/Press-Release-Premier-Campy.pdf. (accessed 12 April 2015).

Bodrossy, L. and A. Sessitsch. 2004. Oligonucleotide microarrays in microbial diagnostics. Curr. Opin. Microbiol. 7(3): 245–54.

Busse, H.J., E.B. Denner and W. Lubitz. 1996. Classification and identification of bacteria: Current approaches to an old problem. Overview of methods used in bacterial systematics. J. Biotechnol. 47(1): 3–38.

Byrne, L., R. Elson, T.J. Dallman, N. Perry, P. Ashton, J. Wain, G.K. Adak, K.A. Grant and C. Jenkins. 2014. Evaluating the Use of Multilocus Variable Number Tandem Repeat Analysis of Shiga Toxin-Producing *Escherichia coli* O157 as a Routine Public Health Tool in England. PLoS One. 9(1): e85901.

Chan, J.Z., M.J. Pallen, B. Oppenheim and C. Constantinidou. 2012. Genome sequencing in clinical microbiology. Nat. Biotechnol. 30(11): 1068–71.

Chen, K.W., H.J. Lo, Y.H. Lin and S.Y. Li. 2005. Comparison of four molecular typing methods to assess genetic relatedness of Candida Albicans clinical isolates in Taiwan. J. Med. Microbiol. 54(3): 249–58.

Chung, M., H. de Lencastre, P. Matthews, A. Tomasz, I. Adamsson, M. Aires de Sousa, T. Camou, C. Cocuzza, A. Corso, I. Couto, A. Dominguez, M. Gniadkowski, R. Goering, A. Gomes, K. Kikuchi, A. Marchese, R. Mato, O. Melter, D. Oliveira, R. Palacio, R. Sa-Leao, I. Santos Sanches, J.H. Song, P.T. Tassios and P. Villari. 2000. Molecular typing of methicillin-resistant *Staphylococcus aureus* by pulsed-field gel electrophoresis: comparison of results obtained in a multilaboratory effort using identical protocols and mrsa strains. Microb. Drug Resist. 6(3): 189–98.

Clark, Clifford G., Tamara M.A.C. Kruk, Louis Bryden, Yolanda Hirvi, Rafiq Ahmed and Frank G. Rodgers. 2003. Subtyping of *Salmonella enterica* serotype enteritidis strains by manual and automated psti-sphi ribotyping. Journal of Clinical Microbiology. 41(1): 27–33.

Cronquist, A.B., R.K. Mody, R. Atkinson, J. Besser, M. Tobin D'Angelo, S. Hurd, T. Robinson, C. Nicholson and B.E. Mahon. 2012. Impacts of culture-independent diagnostic practices on public health surveillance for bacterial enteric pathogens. Clin. Infect. Dis. 54 Suppl. 5: S432–9.

Cunningham, S.A., L.M. Sloan, L.M. Nyre, E.A. Vetter, J. Mandrekar and R. Patel. 2010. Three-hour molecular detection of Campylobacter, Salmonella, Yersinia, and Shigella species in feces with accuracy as high as that of culture. J. Clin. Microbiol. 48(8): 2929–33.

Davis, M.A., D.D. Hancock, T.E. Besser and D.R. Call. 2003. Evaluation of pulsed-field gel electrophoresis as a tool for determining the degree of genetic relatedness between strains of *Escherichia coli* O157:H7. J. Clin. Microbiol. 41(5): 1843–9.

Deng, X., N. Shariat, E.M. Driebe, C.C. Roe, B. Tolar, E. Trees, P. Keim, W. Zhang, E.G. Dudley, P.I. Fields and D.M. Engelthaler. 2015. Comparative analysis of subtyping Methods against a whole-genome-sequencing standard for Salmonella enterica serotype enteritidis. J. Clin. Microbiol. 53(1): 212–8.

Drancourt, M., V. Roux, L.V. Dang, L. Tran-Hung, D. Castex, V. Chenal-Francisque, H. Ogata, P.E. Fournier, E. Crubezy and D. Raoult. 2004. Genotyping, orientalis-like Yersinia pestis, and plague pandemics. Emerg. Infect. Dis. 10(9): 1585–92.

EFSA BIOHAZ Panel EFSA Panel on Biological. 2013. Scientific opinion on the evaluation of molecular typing methods for major food-borne microbiological hazards and their use for attribution modelling, outbreak investigation and scanning surveillance: Part 1 (Evaluation of Methods and Applications). EFSA J. 11: 3502–3586.

Eom, H.S., B.H. Hwang, D.H. Kim, I.B. Lee, Y.H. Kim and H.J. Cha. 2007. Multiple detection of food-borne pathogenic bacteria using a novel 16s rDNA-based oligonucleotide signature Chip. Biosens. Bioelectron. 22(6): 845–53.

Eyre, D.W., M.L. Cule, D.J. Wilson, D. Griffiths, A. Vaughan, L. O'Connor, C.L. Ip, T. Golubchik, E.M. Batty, J.M. Finney, D.H. Wyllie, X. Didelot, P. Piazza, R. Bowden, K.E. Dingle, R.M. Harding, D.W. Crook, M.H. Wilcox, T.E. Peto and A.S. Walker. 2013. Diverse sources of *C. difficile* infection identified on whole-genome sequencing. N. Engl. J. Med. 369: 1195–1205.

Field, D., J. Hughes and E.R. Moxon. 2004. Using the genome to understand pathogenicity. Methods Mol. Biol. 266: 261–87.

Fournier, P.E., Y. Zhu, H. Ogata and D. Raoult. 2004. Use of highly variable intergenic spacer sequences for multispacer typing of rickettsia conorii strains. J. Clin. Microbiol. 42(12): 5757–66.

Frenay, H.M., J.P. Theelen, L.M. Schouls, C.M. Vandenbroucke-Grauls, J. Verhoef, W.J. van Leeuwen and F.R. Mooi. 1994. Discrimination of epidemic and nonepidemic methicillin-resistant *Staphylococcus aureus* strains on the basis of protein a gene polymorphism. J. Clin. Microbiol. 32(3): 846–7.

Hahm, B.K., Y. Maldonado, E. Schreiber, A.K. Bhunia and C.H. Nakatsu. 2003. Subtyping of foodborne and environmental isolates of *Escherichia coli* by multiplex-PCR, REP-PCR, PFGE, Ribotyping and Aflp. J. Microbiol. Methods. 53(3): 387–99.

Hoefer, D., S. Hurd, C. Medus, A. Cronquist, S. Hanna, J. Hatch, T. Hayes, K. Larson, C. Nicholson, K. Wymore, M. Tobin-D'Angelo, N. Strockbine, P. Snippes, R. Atkinson, P. M. Griffin and L.H. Gould. 2011. Laboratory practices for the identification of shiga toxin-producing *Escherichia coli* in the United States, Foodnet Sites, 2007. Foodborne Pathog. Dis. 8(4): 555–60.

Hyytia-Trees, E., S.C. Smole, P.A. Fields, B. Swaminathan and E.M. Ribot. 2006. Second generation subtyping: a proposed pulsenet protocol for Multiple-Locus Variable-Number Tandem Repeat Analysis of shiga toxin-producing *Escherichia coli* O157 (Stec O157). Foodborne Pathog. Dis. 3(1): 118–31.

Jain, S., S.A. Bidol, J.L. Austin, E. Berl, F. Elson, M. Lemaile-Williams, M. Deasy, 3rd, M.E. Moll, V. Rea, J.D. Vojdani, P.A. Yu, R.M. Hoekstra, C.R. Braden and M.F. Lynch. 2009. Multistate outbreak of Salmonella Typhimurium and saintpaul infections associated with unpasteurized orange juice—United States, 2005. Clin. Infect. Dis. 48(8): 1065–71.

Keim, P., L.B. Price, A.M. Klevytska, K.L. Smith, J.M. Schupp, R. Okinaka, P.J. Jackson and M.E. Hugh-Jones. 2000. Multiple-locus variable-number tandem repeat analysis reveals genetic relationships within Bacillus anthracis. J. Bacteriol. 182(10): 2928–36.

Kim, H.J., S.H. Park, T.H. Lee, B.H. Nahm, Y.R. Kim and H.Y. Kim. 2008. Microarray detection of food-borne pathogens using specific probes prepared by comparative genomics. Biosens. Bioelectron. 24(2): 238–46.

Korczak, B.M., M. Zurfluh, S. Emler, J. Kuhn-Oertli and P. Kuhnert. 2009. Multiplex strategy for multilocus sequence typing, fla typing, and genetic determination of antimicrobial resistance of Campylobacter jejuni and Campylobacter coli isolates collected in Switzerland. J. Clin. Microbiol. 47(7): 1996–2007.

Lambert, D., C.D. Carrillo, A.G. Koziol, P. Manninger and B.W. Blais. 2015. Genesippr: A Rapid Whole-Genome Approach for the Identification and Characterization of Foodborne Pathogens Such as Priority Shiga Toxigenic *Escherichia coli*. PLoS One. 10(4): e0122928.

Larsson, J.T., M. Torpdahl and E. Moller Nielsen. 2013. Proof-of-concept study for successful inter-laboratory comparison of Mlva results. Euro. Surveill. 18(35): 20566.

Larsson, J.T., M. Torpdahl, R.F. Petersen, G. Sorensen, B.A. Lindstedt and E.M. Nielsen. 2009. Development of a new nomenclature for Salmonella Typhimurium multilocus variable number of tandem repeats analysis (Mlva). Euro Surveill. 14(15).

Li, W., B.B. Chomel, S. Maruyama, L. Guptil, A. Sander, D. Raoult and P.E. Fournier. 2006. Multispacer typing to study the genotypic distribution of bartonella henselae populations. J. Clin. Microbiol. 44(7): 2499–506.

Li, W., D. Raoult and P.E. Fournier. 2009. Bacterial strain typing in the genomic era. FEMS Microbiol. Rev. 33(5): 892–916.

Lienau, E.K., E. Strain, C. Wang, J. Zheng, A.R. Ottesen, C.E. Keys, T.S. Hammack, S.M. Musser, E.W. Brown, M.W. Allard, G. Cao, J. Meng and R. Stones. 2011. Identification of a Salmonellosis outbreak by means of molecular sequencing. N. Engl. J. Med. 364(10): 981–2.

Lim, H., K.H. Lee, C.H. Hong, G.J. Bahk and W.S. Choi. 2005. Comparison of four molecular typing methods for the differentiation of *Salmonella* spp. Int. J. Food Microbiol. 105(3): 411–8.

Lindstedt, B.A. 2005. Multiple-locus variable number tandem repeats analysis for genetic fingerprinting of pathogenic bacteria. Electrophoresis. 26(13): 2567–82.

Lo, Y.M. and K.C. Chan. 2006. Introduction to the polymerase chain reaction. Methods Mol. Biol. 336: 1–10.

Loman, N.J., C. Constantinidou, J.Z. Chan, M. Halachev, M. Sergeant, C.W. Penn, E.R. Robinson and M.J. Pallen. 2012. High-throughput bacterial genome sequencing: an embarrassment of choice, a world of opportunity. Nat. Rev. Microbiol. 10(9): 599–606.

Lupski, J.R. and G.M. Weinstock. 2000. Short, Interspersed repetitive DNA sequences in prokaryotic genomes. J. Bacteriol. 174(14): 4525–9.

Maiden, M.C., J.A. Bygraves, E. Feil, G. Morelli, J.E. Russell, R. Urwin, Q. Zhang, J. Zhou, K. Zurth, D.A. Caugant, I.M. Feavers, M. Achtman and B.G. Spratt. 1998. Multilocus sequence typing: a portable approach to the identification of clones within populations of pathogenic microorganisms. Proc. Natl. Acad. Sci. USA. 95(6): 3140–5.

Martin, I.E., S.D. Tyler, K.D. Tyler, R. Khakhria and W.M. Johnson. 1996. Evaluation of ribotyping as epidemiologic tool for typing *Escherichia coli* serogroup O157 isolates. Journal of Clinical Microbiology. 34(3): 720–723.

Mazuet, C., J. Sautereau, C. Legeay, C. Bouchier, P. Bouvet and M.R. Popoff. 2015. An atypical outbreak of food-borne botulism due to *Clostridium botulinum* types B and E from Ham. J. Clin. Microbiol. 53(2): 722–6.

McCollum, J.T., A.B. Cronquist, B.J. Silk, K.A. Jackson, K.A. O'Connor, S. Cosgrove, J.P. Gossack, S.S. Parachini, N.S. Jain, P. Ettestad, M. Ibraheem, V. Cantu, M. Joshi, T. DuVernoy, N.W. Fogg, Jr., J.R. Gorny, K.M. Mogen, C. Spires, P. Teitell, L.A. Joseph, C.L. Tarr, M. Imanishi, K.P. Neil, R.V. Tauxe and B.E. Mahon. 2013. Multistate outbreak of listeriosis associated with Cantaloupe. N. Engl. J. Med. 369(10): 944–53.

Meinersmann, R.J., L.O. Helsel, P.I. Fields and K.L. Hiett. 1997. Discrimination of *Campylobacter jejuni* isolates by fla gene sequencing. J. Clin. Microbiol. 35(11): 2810–4.

Miko, A., B.A. Lindstedt, L.T. Brandal, I. Lobersli and L. Beutin. 2010. Evaluation of multiple-locus variable number of tandem-repeats analysis (Mlva) as a method for identification of clonal groups among Enteropathogenic, Enterohaemorrhagic and Avirulent *Escherichia coli* O26 strains. FEMS Microbiol. Lett. 303(2): 137–46.

Noller, A.C., M.C. McEllistrem, A.G. Pacheco, D.J. Boxrud and L.H. Harrison. 2003. Multilocus variable-number tandem repeat analysis distinguishes outbreak and sporadic *Escherichia coli* O157:H7 Isolates. J. Clin. Microbiol. 41(12): 5389–97.

Prevention, Centers for Disease Control and—"Two Multistate Outbreaks of Shiga Toxin-Producing *Escherichia coli* Infections Linked to Beef from a Single Slaughter Facility—United States, 2008". 2010. MMWR Morb. Mortal Wkly Rep. 59(18): 557–60.

Price, L.B., M. Stegger, H. Hasman, M. Aziz, J. Larsen, P.S. Andersen, T. Pearson, A.E. Waters, J.T. Foster, J. Schupp, J. Gillece, E. Driebe, C.M. Liu, B. Springer, I. Zdovc, A. Battisti, A. Franco, J. Zmudzki, S. Schwarz, P. Butaye, E. Jouy, C. Pomba, M.C. Porrero, R. Ruimy, T.C. Smith, D.A. Robinson, J.S. Weese, C.S. Arriola, F. Yu, F. Laurent, P. Keim, R. Skov and F.M. Aarestrup. 2012. *Staphylococcus aureus* Cc398: Host Adaptation and Emergence of Methicillin Resistance in Livestock. MBio. 3(1).

Revazishvili, T., M. Kotetishvili, O.C. Stine, A.S. Kreger, J.G. Morris, Jr. and A. Sulakvelidze. 2004. Comparative analysis of multilocus sequence typing and pulsed-field gel electrophoresis for characterizing *Listeria monocytogenes* strains isolated from environmental and clinical sources. J. Clin. Microbiol. 42(1): 276–85.

Ribot, E.M., M.A. Fair, R. Gautom, D.N. Cameron, S.B. Hunter, B. Swaminathan and T.J. Barrett. 2006. Standardization of pulsed-field gel electrophoresis protocols for the subtyping of *Escherichia coli* O157:H7, Salmonella, and Shigella for Pulsenet. Foodborne Pathog. Dis. 3(1): 59–67.

Ruiz-Garbajosa, P., M.J. Bonten, D.A. Robinson, J. Top, S.R. Nallapareddy, C. Torres, T.M. Coque, R. Canton, F. Baquero, B.E. Murray, R. del Campo and R.J. Willems. 2006. Multilocus sequence typing scheme for *Enterococcus faecalis* reveals hospital-adapted genetic complexes in a background of high rates of recombination. J. Clin. Microbiol. 44(6): 2220–8.

Sanger, F., S. Nicklen and A.R. Coulson. 1977. DNA sequencing with chain-terminating inhibitors. Proc. Natl. Acad. Sci. USA. 74(12): 5463–7.

Scheutz, F., L.D. Teel, L. Beutin, D. Pierard, G. Buvens, H. Karch, A. Mellmann, A. Caprioli, R. Tozzoli, S. Morabito, N.A. Strockbine, A.R. Melton-Celsa, M. Sanchez, S. Persson and A.D. O'Brien. 2012. Multicenter evaluation of a sequence-based protocol for subtyping shiga toxins and standardizing Stx nomenclature. J. Clin. Microbiol. 50(9): 2951–63.

Schwartz, D.C. and C.R. Cantor. 1984. Separation of Yeast Chromosome-Sized Dnas by Pulsed Field Gradient Gel Electrophoresis. Cell. 37(1): 67–75.

Sergeev, N., M. Distler, S. Courtney, S.F. Al-Khaldi, D. Volokhov, V. Chizhikov and A. Rasooly. 2004. Multipathogen oligonucleotide microarray for environmental and biodefense applications. Biosens. Bioelectron. 20(4): 684–98.

Swaminathan, B., T.J. Barrett, S.B. Hunter and R.V. Tauxe. 2001. Pulsenet: The molecular subtyping network for foodborne bacterial disease surveillance, United States. Emerg. Infect. Dis. 7(3): 382–9.

Tenover, F.C., R.R. Vaughn, L.K. McDougal, G.E. Fosheim and J.E. McGowan, Jr. 2007. Multiple-locus variable-number tandem-repeat assay analysis of methicillin-resistant *Staphylococcus aureus* Strains. J. Clin. Microbiol. 45(7): 2215–9.

Thibodeau, S.N. 1987. Use of restriction fragment length polymorphism analysis for detecting carriers of "Fragile X" syndrome. Clin. Chem. 33(10): 1726–30.

van Gent, M., M.J. Bart, H.G. van der Heide, K.J. Heuvelman, T. Kallonen, Q. He, J. Mertsola, A. Advani, H.O. Hallander, K. Janssens, P.W. Hermans and F.R. Mooi. 2011. Snp-based typing: a useful tool to study Bordetella pertussis populations. PLoS One. 6(5): e20340.

Wirth, T., D. Falush, R. Lan, F. Colles, P. Mensa, L.H. Wieler, H. Karch, P.R. Reeves, M.C. Maiden, H. Ochman and M. Achtman. 2006. Sex and virulence in *Escherichia coli*: an evolutionary perspective. Mol. Microbiol. 60(5): 1136–51.

CHAPTER 7

Diagnosis and Management of Foodborne Infectious Diseases

Ted Steiner

Introduction

Lack of access to clean water for drinking, bathing, washing and food preparation underlies much of the childhood mortality, developmental delays and poverty that plague the developing world. It is no surprise that intestinal infections related to contaminated food and water are a leading cause of childhood mortality in these areas and the long term consequences in children who survive are profound (Guerrant et al. 2013). However, even in developed countries that can afford high standards of sanitation, diarrheal infections are extremely common, second only to upper respiratory infections as a cause of illness in the population as a whole (Guerrant et al. 2001). While some of these infections are acquired from direct person-to-person spread, the majority are largely or even exclusively acquired through ingestion of contaminated food or water.

The list of food and waterborne infectious agents is somewhat long, and includes a number of common and rare pathogens. These can be divided somewhat arbitrarily along different categories, depending on the type of illness they cause, the primary source of the contamination, the type of food generally implicated, or the severity of infections. Some examples of these categories are shown in Table 1.

Associate Professor of Medicine, Division of Infectious Diseases, University of British Columbia, Vancouver, BC, Canada.
Email: tsteiner@mail.ubc.ca

Type of illness	Watery diarrhea	Inflammatory diarrhea	Nausea and vomiting	Systemic illness
examples	Cholera, enteropathogenic *E. coli*	Nontyphoidal *Salmonella*, *Campylobacter*	Norovirus, *Bacillus cereus* toxin	*Salmonella* Typhi, *Listeria*
Source of contamination	Human-to-human through food preparation	Colonization of livestock or fish	Contamination of food and water sources	Improper food storage
examples	*Salmonella* Typhi, *Shigella*	*Campylobacter*, *Vibrio vulnificus*	Shigatoxigenic *E. coli*, *Cryptosporidium*	*B. cereus*, *Staphylococcus aureus*
Type of food implicated	Poultry	Meat	Dairy	Water
examples	*Campylobacter*, *Salmonella*	Enterohemorrhagic *E. coli*, *Yersinia enterocolitica*	*Listeria, Brucella*	*Giardia*, *Cryptosporidium*, *dracunculiasis*
Severity	Brief, self-limited	Self-limited but prolonged or severe	Frequently severe or fatal if untreated	Highly lethal if untreated
examples	anisikiasis, norovirus	*Camplyobacter*, *Giardia*	*Salmonella* Typhi, *Listeria*	Botulism, cholera

Foodborne infectious diseases include those in which an infectious agent (prion, virus, bacteria, protozoon, etc.) is ingested along with food and replicates within the host to cause disease. This stands in contrast to important foodborne diseases that are mediated by ingestion of pre-formed toxins, even those that have a microbial origin (e.g., botulism, ciguatera), since there is no actual infection occurring. This review will focus on the true infectious causes of foodborne illness.

General Principles of Diagnosis

A foodborne illness should be suspected in patients who develop relative acute onset of gastrointestinal symptoms, including nausea, vomiting, diarrhea, abdominal pain, cramping, or blood in stools. It is seldom possible to identify the specific infectious agent causing a foodborne illness on the basis of clinical presentation alone. Instead, most clinicians rely on the patient's history (e.g., symptoms, exposure and timing) to narrow the disease down to a syndrome, which in turn leads to appropriate diagnostic algorithms. In many cases, this approach will negate the need for any diagnostic testing, since the self-limited or untreatable nature of the infection will be obvious.

These clinical syndromes of infectious foodborne illness include:

- **"Food poisoning" through toxin ingestion.** While nausea and vomiting can occur with any intestinal infection, when they are the first and/or the major symptoms this usually reflects primary involvement of the proximal small intestine or, less commonly, the stomach. There are two main scenarios in which this occurs. The first is through ingestion of preformed toxins that have contaminated food during its preparation and/or storage. This included toxins made by bacteria during improper food storage. These toxins may be heat stable and survive gastric acid to reach their target sites within the gut. Intoxication then leads to abrupt onset of severe nausea and vomiting, sometimes followed by abdominal pain, cramps and watery diarrhea, usually without fever. Symptoms generally resolve within 24 h (Chan et al. 1994). In some cases (such as with non-emetic *B. cereus*), cramping and diarrhea predominate (Checkley et al. 1996). Patients may report a common source illness with other patients (such as sharing potato salad at the same picnic). The major toxigenic organisms in these cases are *Staphylococcus aureus*, *Bacillus cereus*, and *Clostridium perfringens*. Commonly implicated foods are prepared dishes (salads, rice dishes, etc.) that are not kept adequately refrigerated.

    Because these illnesses are self limited within a time frame of hours, there is no indication for a diagnostic workup. Moreover, the toxins are difficult to detect in samples from patients, but are more commonly identified in the suspected source food, which is indicated only in outbreak situations, using a variety of molecular techniques (Chiao et al. 2013).

- **Viral gastroenteritis.** The other major cause of a vomiting-predominant illness is viral infection of the upper small intestine. These viruses are generally extremely infectious and can be spread through food handled by a person shedding virus. The most common of these in the adult population is norovirus, which typically presents with abrupt onset of nausea and vomiting followed by watery diarrhea lasting 2–3 days (Al Mamun et al. 1996). Rotavirus, which is most common in children, tends to produce a more severe and prolonged illness with less vomiting (Choi et al. 2004).

- **Watery diarrheal infection.** The human gut secretes and absorbs as much as 8–10 L of fluid a day through carefully coordinated hormonally and neurally controlled ionic channels. Many infectious agents can colonize the small intestine and disrupt this fluid balance, through toxin secretion, inflammation, or both. Non-toxigenic organisms (such as enteropathogenic *E. coli*, *Giardia*, *Cryptosporidium*, and *Cyclospora*) colonize the small intestinal epithelium, with a resulting inflammatory response that causes fluid hypersecretion and impaired absorption due to blunting of the villi. This typically leads to abdominal cramping, bloating, flatulence, and watery, loose, or occasionally greasy stools. In contrast, toxigenic bacteria such as *Vibrio cholerae* and enterotoxigenic *E. coli* are less inflammatory, but cause fluid hypersecretion due to direct effects of their toxins on the intestinal epithelium. Infected patients typically present with acute onset of high-volume, watery diarrhea, sometimes accompanied by nausea, vomiting, and low-grade fever.

- **Inflammatory diarrhea.** A number of pathogens possess specific virulence traits that allow them invade intestinal epithelial cells. This invasion, which occurs typically in the distal small bowel or colon, frequently leads to an exuberant local inflammatory response that causes further tissue destruction and release of inflammatory cytokines. As a result, patients experience diarrhea (which may be bloody or purulent), often associated with pain or cramps, high fever and malaise. Nausea and vomiting may be present but are typically less prominent. In some cases, patients pass frequent, low-volume stools with tenesmus (urgency that is not relieved with defecation), which are together characteristic of dysentery. Inflammatory diarrhea can usually be distinguished from viral gastroenteritis by the presence of a high fever, blood in stools, or duration more than 48 h.

- **Hemorrhagic colitis.** This is one of the few diarrheal syndromes that is associated with a single pathogen type (shigatoxigenic *E. coli*). It classically presents acutely with watery diarrhea, abdominal cramps, and tenderness, progressing quickly to frankly bloody diarrhea and severe abdominal pain. Fever is conspicuously low or absent in most cases, which distinguishes this condition from inflammatory enterocolitis (Dunne et al. 2003, Cooke 1968). While the intestinal symptoms are self-limited, hemorrhagic colitis can progress to hemolytic-uremic syndrome with potentially fatal outcomes. It is for this reason that diagnostic testing is generally recommended for all patients presenting with bloody diarrhea (Guerrant et al. 2001).

- **Enteric fever.** *Salmonella enterica* serovar Typhi (*S.* Typhi) and related strains can be spread directly from person to person or transmitted through food and water after contamination by infected individuals. "Typhoid Mary", a chronic S. Typhi carrier, is believed to have infected 51 people through her work as a cook, demonstrating the dangers of transmission. Typhoid fever typically presents with fever, abdominal discomfort, headache, and other nonspecific symptoms (Coopersmith et al. 2002). These symptoms tend to progress in a stepwise fashion, becoming more severe until treated or they become fatal. Diarrhea is not a typical symptom despite the intestinal origin of infection.

- **Other pathogen-specific conditions.** Some foodborne pathogens have unique life cycles and disease characteristics that do not fit into the above categories. For example, *Listeria monocytogenes* causes an invasive illness that can begin with nonspecific diarrhea, but then progresses to bacteremia, meningitis, or other systemic illness. *Yersinia enterocolitica* can also produce a nonspecific diarrheal illness, but is recognized as a cause of "pseudoappendicitis" (severe right lower quadrant pain) due to involvement of mesenteric lymph nodes. *Toxoplasma gondii* is a protozoon that generally causes a benign or asymptomatic, self-limited infection after ingestion of contaminated water or food, but in immune compromised hosts it can cause a severe or fatal encephalitis and can cause fetal anomalies if acquired during pregnancy.

Diagnostic Approaches to Foodborne Illness

The Infectious Diseases Society of America has produced guidelines for the management of infectious diarrhea (Guerrant et al. 2001). These guidelines were first published in 2001, and a revised version is due to be out in 2015. They provide recommendations regarding when to consider ordering diagnostic testing to evaluate foodborne illness and which are the tests recommended. Some of the recommendations from the 2001 guidelines are likely to be revised, given improved technology, particularly the ability to apply molecular diagnostic methods to routine clinical investigations. For example, electron microscopy and viral culture of stool have been all but supplanted by PCR to diagnose viral gastroenteritis. However, several important principles remain. The most important is that the presenting syndrome of illness should guide the appropriate use of diagnostic testing, as follows:

1. *Acute food poisoning.* Since these illnesses are self-limited, analysis of vomitus, stool, or blood is not generally recommended.
2. *Viral gastroenteritis.* Again, given the self-limited nature of these infections, diagnostic testing is not generally required or recommended. However, in an outbreak setting, identification of the offending virus may be helpful in guiding infection control practices or ruling out other agents. PCR on stool is considered the current gold standard for norovirus testing, although commercial immunoassays on stool are also available (Fisman et al. 2009). Stool antigen testing is recommended for rotavirus, although PCR and EIA methods are also available (Anonymous c).
3. *Watery diarrhea.* In most cases, diagnostic testing for watery diarrhea without fever or blood is not recommended, due to the self-limited

nature of the infections. However, there are situations where testing is recommended. This includes prolonged illness (> 14 d), immunocompromised hosts, or known contacts or exposures of concern (for example recent travel to a cholera-affected area). Most cases of travel-related diarrhea do not require diagnostic evaluation, but rather empiric antibiotic therapy is recommended based on solid clinical trial evidence.

When diagnostic testing is required, it should include a routine stool culture and examination for protozoa (either by antigen testing using commercial kits, or microscopy in qualified laboratories). In the case of severe immune compromise, acid fast staining (for microsporidia and *Cryptosporidium*), acid fast culture (for *Mycobacterium avium* complex), and PCR for cytomegalovirus should be considered on stool, although endoscopic sampling is often required.

There are important pathotypes of *E. coli* causing watery diarrhea that cannot readily be detected by routine culture. These include enteroaggregative (EAEC), enterotoxigenic (ETEC), enteropathogenic (EPEC), and diffusely adherent (DAEC) strains. Molecular tests are available to identify these organisms, but they are currently used only as research tools. It is possible that within the near future these tests will be more widely available for clinical use. The one exception is molecular testing for STEC (to detect shigatoxins), which is an available method to diagnose hemorrhagic colitis (see below).

4. *Inflammatory or bloody diarrhea.* Most healthy patients who present with inflammatory diarrhea will have a self-limited illness, so routine culture of stool in all such cases is not necessary. Moreover, some of these illnesses are either not improved (e.g., *Salmonella*) or minimally improved (e.g., *Campylobacter*) with antibiotic treatment. Despite this, most authorities recommend performing diagnostic testing in patients with obvious inflammatory diarrhea, particularly those with the most severe presentations (high fever, multiple stools per day, significant volume depletion, etc.), not only to guide potential therapy, but to help identify the source of infection and potential outbreak situations. In these cases, a routine stool culture is the most useful test. However, there are specific circumstances where additional testing is recommended. Recent foreign travel or high risk sexual contacts (e.g., men who have sex with men) should prompt testing for *Entamoeba histolytica* using microscopy or antigen testing. Any patient with a history of recent abdominal surgery, antibiotic use, or cancer chemotherapy should have testing for *C. difficile*. When enteric fever, invasive *Salmonella* infection, or listeriosis are suspected, blood cultures should be obtained. Finally, in locations where routine testing for STEC on all stool cultures is not performed, it should be specifically requested in any patient with bloody diarrhea.

The specific methods employed in stool testing vary from laboratory to laboratory, but in general include the following:

1. *Stool culture*—This is performed by plating stool samples on a variety of culture plates that allow selective growth and identification of the major bacterial pathogens. Culture conditions should include microaerophilic and temperature variations to allow for enrichment and ideal growth of specific pathogens (as described below). Antibiotic susceptibility testing of isolated pathogens is generally recommended given increasing resistance to formerly first-line antibiotics.
2. *Microscopy*—This is generally performed on formalin-fixed stool samples collected in special vials to prevent degradation. Some pathogens and their eggs are easily detected under light microscopy; others require acid-fast staining or fluorescent microscopy after antibody staining.
3. *Antigen testing*—commercially available EIAs are available for several pathogens, and in general have the advantage of requiring less expertise than microscopy. In addition, antigen testing can help to distinguish pathogenic from nonpathogenic organisms that share similar appearance under microscopy (such as *Entamoeba histolytica* versus *Entamoeba dispar*). Finally, rapid antigen assays are frequently used in the diagnosis of *C. difficile* to detect secreted toxins as well as bacterial antigens.
4. *Molecular testing*—identification of characteristic pathogenic genes in bacteria, or of viral nucleic acids, can be readily performed using commercial or, in some cases, "home-grown" assays. These include gene probe kits, PCR, and, more frequently nowadays, real-time quantitative PCR. These tests are frequently used to detect intestinal viruses, STEC, and *C. difficile.*
5. *Specialized testing*—some laboratories still perform mammalian cell culture for clinical diagnosis. This requires considerable expertise, but can still be useful in the diagnosis of STEC, *C. difficile*, and viral infections. Older tests using live animals to identify invasive organisms or diarrheagenic toxins are no long used in clinical practice.

Approach to Diagnosis and Management of Specific Foodborne Infections

The following section discusses the major foodborne infections, with a brief summary of pathogenesis, diagnosis and treatment. There are a number of rare pathogens identified in foodborne gastroenteritis that are not discussed in the interest of space. Organisms are classified by taxonomy for the sake of simplicity.

Prions. Prions are naturally occurring proteins that have undergone a conformational shift that alters their function, leading to accumulation within cells. Entry of prions into the central nervous system leads to a progressive, fatal encephalopathy. Genetic and sporadic prion diseases have been described, and the one of greatest concern in western countries is variant Creutzfeld-Jacob disease (vCJD) (Anonymous d) Like its sporadic counterpart (CJD), it is a progressive, untreatable degenerative illness affecting diffuse areas of the brain. It typically presents with progressive psychiatric and neurologic symptoms. vCJD tends to occur in younger patients and has a slower progression than CJD. Diagnosis can be difficult, requiring tissue demonstration of prions or characteristic cellular inclusions for confirmation. There is no treatment shown to be effective and the median survival is 14 months.

The foodborne nature of vCJD remains unproven, but it is believed to arise following consumption of meat from cows carrying the prion, who develop a disease known as bovine spongiform encephalopathy (BSE). A spike in cases of vCJD in the UK beginning in 1996 followed a rise in BSE, and subsequent bans on feeding animal products to cattle have lead to a steady decline. Transmission of the prion from cows to sheep has been demonstrated experimentally, and transfusion-associated cases in humans are also suspected. Ongoing surveillance for BSE and removal of suspected animals from the human food chain, along with bans on feeding animal products to cows are the best ways to avoid this illness.

Viruses. The strong species specificity of most viruses means that human intestinal viruses are spread from person to person. However, their high infectivity makes food and waterborne transmission common, following contamination by infectious fluids (usually stool or vomitus) from infected people. Hence, they are clearly important causes of foodborne illness. There are four major human intestinal viruses: norovirus, rotavirus, adenovirus, and astrovirus. All produce a self-limited illness and there is no effective antiviral treatment. Supportive care with fluid resuscitation (orally if possible) and antiemetics is the mainstay of therapy. When indicated, testing of stool using antigen testing of PCR provides the best means of diagnosis.

- **Norovirus.** A member of the caliciviridae family, norovirus was formerly known as Norwalk-like virus following an outbreak in Norwalk, Ohio. A related calicivirus, sapovirus, produces a similar illness. Norovirus is the most common cause of gastroenteritis in the US, with an estimated 19–21 million cases per year (Hall et al. 2013). Despite this high incidence, fatalaties are rare (fewer than 1000 per year). The virus is highly infectious, with as few as 18, but more likely just over 1000 virions needed to cause disease (Atmar et al. 2014). After a brief incubation period of 12–72 h, illness begins abruptly with nausea, vomiting, and/or diarrhea. Vomiting is typically the predominant symptom, and the illness lasts 48–72 hours. Patients may shed virus for weeks, although those who are symptomatic are at highest risk of transmission. While a strong antibody response develops in most patients, protection is not long-lived, and repeat attacks are common.

- **Rotavirus.** Like norovirus, rotavirus is an extremely common infection. Antibody surveys suggest that in the pre-vaccination era, almost all children were infected before adulthood. Lasting protection does develop following natural infection in most people, although re-infections can occur in adults, particularly with immune compromise. The illness has seasonal peaks during the winter months and those most heavily affected are infants and very young children. The major symptom is diarrhea, which tends to be more prolonged and severe than norovirus, but remains noninflammatory. Vomiting, dehydration and fever are also common. The average duration of illness is 8 days (Gurwith et al. 1981). Treatment is supportive and fluid restoration is the cornerstone of therapy. Studies have demonstrated benefit of adjunctive treatments including anti-emetics (5-HT antagonists in particular) and probiotics (Guarino et al. 2014, Freedman 2007). Zinc supplementation should be considered in children who are malnourished or living in impoverished areas with low dietary zinc. Rotavirus can be effectively prevented by vaccination which is currently recommended for all infants beginning at two months of age (Anonymous c).

- **Enteric adenovirus.** There are numerous serotypes of adenovirus that cause a variety of diseases, including respiratory infections (upper and lower), conjunctivitis, hemorrhagic cystitis, and disseminated, fatal infection in severely immunocompromised people. Subgroup F (serotypes 40 and 41) are associated with gastroenteritis in children, with no specific characteristic features (Krajden et al. 1990).

- **Astrovirus.** Astroviruses cause sporadic childhood diarrheal illness, and can be spread through food and water or in an outbreak setting (Appleton 1987). The illness they cause has no particular distinguishing characteristics and is generally considered mild (Gabbay et al. 2007) although in immunocompromised hosts it can be severe and prolonged.

Bacteria

- **Salmonella.** *Salmonella enterica* is the species designation for human pathogenic strains, and includes a number of important subspecies or serovars. Of those, serovars, Typhi and Paratyphi cause enteric fever, while other common serovars, including Enteritidis, Typhimurium,

Dublinensis, etc. predominantly cause inflammatory gastroenteritis. Humans are an important reservoir of infection, and can easily spread the organism through cooking. In addition, commercial poultry frequently have contamination with *Salmonella*, making consumption of undercooked poultry or raw eggs risk factors (Morgan et al. 2007, White et al. 2007). *Salmonella* gastroenteritis is caused by a coordinated attack involving multiple virulence traits, that allow for adherence, invasion into epithelial cells, and survival within specialized vacuoles in macrophages. However, in the presence of a competent immune system, it is usually a self-limited illness, and treatment with antibiotics provides no overall benefit in healthy adults and indeed increases the likelihood of shedding at 30 days (Bhan et al. 1989). Some experts still recommend treatment for the most severe cases in otherwise healthy adults and most young children and the elderly, who are at higher risk of complications. In addition, patients with immune compromise (AIDS, solid organ transplants, etc.) are at risk of severe or fatal outcomes, and even though treatment has never been studied formally in such patients, antibiotic treatment is recommended for them (reviewed in (Bhatnagar et al. 1993)).

- **Typhoid fever.** More properly termed "enteric fever", this condition is common in developing areas and occasionally travelers to these areas. Humans are the only known reservoir, but prolonged shedding is common, and hence contamination of food during preparation is a frequent means of transmission. Typhoid typically begins as a nonspecific febrile illness, due to immune responses to the bacteria as they invade the intestinal epithelium to reside in macrophages, which can disseminate throughout the lymphatic system. This invasion typically causes abdominal pain, sometimes constipation and stepwise progression of fever and prostration. If untreated, the illness can be fatal due to intestinal perforation or sepsis. Typhoid fever was historically difficult to diagnose due to low sensitivity of blood and stool cultures, and serologic diagnosis (e.g., the Widal test) or even bone marrow cultures were frequently employed. However, modern blood culture methods have much better sensitivity, and serology has been shown to lack sensitivity and specificity, so it is no longer recommended (Dutta et al. 2006). Patients with confirmed enteric fever should always receive antibiotic treatment, and follow up stool cultures to confirm cure are recommended in those at risk of spreading typhoid to others (such as food handlers or health care workers) (From: The Indian Society of Critical Care Medicine Tropical Fever Group et al. 2014, Balasegaram et al. 2012).

- **Shigella.** The genus *Shigella* consists of four main pathogenic species, all of which infect humans exclusively. While genetically extremely closely related to *E. coli*, *Shigella* possesses a virulence plasmid that encodes a complement of invasion proteins that allow the bacteria to enter and move between cells, leaving behind necrosis and considerable inflammation. As a result, shigellosis is a typically severe and highly inflammatory gastroenteritis. Indeed, having *Shigella* as opposed to other pathogens is a risk factor for mortality in children hospitalized with diarrhea (Bhatnagar et al. 1998). *Shigella* has a very low infectious dose (possibly as few as 10–100 organisms in nature, and perhaps more in volunteer challenge models (Porter et al. 2013)) and hence is easily spread through food or water contaminated by an infected individual. While shigellosis is self-limited in most patients, antibiotics shorten the duration of illness and bacterial shedding and since *Shigella* spreads so easily from person to person, it is generally recommended that all cases be treated (Bian et al. 2004). Unfortunately, antibiotic resistance has been increasing outside North America (Birkenkamp et al. 2000).

- **Campylobacter jejuni.** *Campylobacter* grows well at 42 C, the body temperature of many birds, and is frequently found in commercial poultry (Sahin et al. 2002). Hence, it is most commonly acquired from consumption of undercooked chicken or turkey. Less is known about its pathogenicity than other common diarrheal pathogens, but it does possess invasive virulent traits and produces an inflammatory enterocolitis. Fortunately, fatalities are extremely rare, although bacteremias do occur, largely in immunocompromised people, with a mortality rate of 15% in those cases (Bouckenooghe et al. 1999). Antibiotic treatment in uncomplicated cases shortens the duration of illness by a day or two, particularly when administered early (Bouckenooghe et al. 2000). However, many patients will already be better or beyond the window of antibiotic benefit by the time the culture result is received. However, many experts treat severe cases and cases in immune compromised hosts. While *Campylobacter* resistance to front line agents is increasing in some areas, most isolates in the US remain sensitive to macrolides and fluoroquinolones (Deng et al. 2004).

- **Diarrheagenic E. coli.** There are 5 major pathotypes of *E. coli* that cause primarily diarrheal illness, although there is some overlap among them. Generally they cannot be distinguished through routine culture and identification methods in clinical laboratories (the exception being EHEC O157:H7, which is usually sorbitol-negative in contrast to most strains). For this reason, apart from STEC and EHEC, they are seldom diagnosed outside of clinical trials.

o *Enterohemorrhagic E. coli (EHEC).* The terms EHEC and STEC (shigatoxigenic *E. coli*) were formerly used interchangeably, recognizing that the key virulence trait that causes hemorrhagic colitis is the shigatoxins (or shiga-like toxins) Stx1 and Stx2 (also referred to as SLT-1 and SLT-2). However, more recently it has become clear that EHEC shares additional virulence traits with enteropathogenic *E. coli* (EPEC) that allow it to adhere and damage the epithelial brush border in a characteristic "attaching and effacing" pattern. This may help explain the prominent disease activity of the prototype strain, O157:H7. The intestinal symptoms of EHEC infection, while severe, are self-limited, but systemic absorption of Stx can cause hemolytic-uremic syndrome (HUS), which is fatal in up to 5% of cases and can lead to permanent neurological or renal sequelae in an additional 25% or so (Spinale et al. 2013). EHEC should be sought in all submitted stools with visible blood, or in patients with a history of bloody diarrhea. O157:H7 can be presumptively identified as non-fermenting colonies on sorbitol-MacConkey agar. In addition, commercial assays for Stx1 and Stx2 are widely used, including EIA and PCR, and should also be used to assist in diagnosis of non-O157 EHEC.

While EHEC O157:H7 remains relatively antibiotic sensitive (Braegger et al. 1992), there is evidence that antibiotics can induce increased toxin release from EHEC, leading to worse outcomes. While some studies suggested that antibiotics for EHEC increased the risk of HUS, these were non-controlled, and a meta-analysis of case series and nonrandomized prospective trials found no pooled increased risk of HUS with antibiotics (Safdar et al. 2002). However, given the absence of any RCT data and no real evidence of benefit, most authorities do not recommend antibiotic treatment for hemorrhagic colitis. In patients with HUS, the C5 complement inhibitor eculizumab was beneficial in a small case series (Eaves-Pyles et al. 2001a) but prospective, randomized data are lacking. Other treatments like plasma exchange also lack convincing evidence of benefit.

o *Enteroaggregative E. coli (EAEC).* EAEC possess unique and pathognomonic virulence traits characterized by aggregative adherence to surfaces and cells in a biofilm. Clinically, EAEC may be difficult to distinguish from other diarrheal infections, although some studies have shown a moderately inflammatory phenotype with mucousy diarrhea (Steiner et al. 1998, Nataro and Steiner 2002). It is a major cause of traveler's diarrhea and endemic childhood diarrhea in developing areas; it has a propensity to cause prolonged illness, particularly in immunocompromised people. It probably has both person-to-person and foodborne transmission. EAEC can be easily identified through biofilm formation or adherence to mammalian cells, as well as molecular testing for the master virulence regulator *aggR*, but these tests are not clinically available. No treatment trials have been performed, and many EAEC strains are resistant to front-line antibiotics.

○ *STEC O104:H4.* This EAEC strain caused a large foodborne outbreak of hemorrhagic colitis in Europe in 2011, and the HUS rates in adults were unusually high at 22% (Brooks et al. 2004). The source was identified as contaminated fenugreek seeds. Almost 4000 people were affected. O104:H4 expresses an extended-spectrum ß-lactamase (ESBL), rendering it resistant to cephalosporins, but it remains susceptible to fluoroquinolones, rifaximin, and azithromycin. Unlike EHEC, it is not clear whether antibiotics induce toxin production and release from this strain (Brown et al. 1993, Burke and Axon 1987). Many patients infected in the outbreak were given antibiotics and no association with worse outcome was reported; in fact, treatment with ciprofloxacin was associated with reduced HUS risk in one case series (Burke et al. 1988). Treatment with azithromycin appears to reduce shedding with no effect on clinical outcome (Dunne and O'Neill 2003). Hence, there is insufficient evidence that antibiotics are either harmful or beneficial in O104:H4 infection. Neither eculizumab nor plasma exchange for patient with HUS were found to be beneficial (Dunne et al. 2003, Eaves-Pyles et al. 2001b).

○ *Enterotoxigenic E. coli (ETEC).* ETEC was one of the first pathotypes of diarrheagenic *E. coli* to be described. Its major virulence traits are secretory toxins, LT and ST, which cause fluid hypersecretion from the small intestinal crypts that can be severe enough to resemble cholera, or produce a milder illness with watery diarrhea, low grade fever, and nausea. It is the leading cause of traveler's diarrhea and also causes endemic childhood diarrhea in developing areas. Because of clear RCT data showing the effectiveness of empiric antibiotic treatment of traveler's diarrhea (DuPont et al. 2009), ETEC is generally not sought specifically and identification requires either molecular testing or more unwieldy cell culture methods.

○ *Enteropathogenic E. coli (EPEC).* EPEC is a major cause of infantile, dehydrating diarrhea in developing areas. It has characteristic attaching-and-effacing adherence to epithelial cells that leads to a mildly inflammatory watery diarrhea. It is self-limited and antibiotic treatment is not generally recommended, but ORS and continued breastfeeding are important to avoid volume depletion.

- **Listeria monocytogenes.** This gram-positive bacillus grows well at 4 degrees so contamination of refrigerated foods as a source of outbreaks is well recognized. This includes deli meats, unpasteurized cheeses, and even pasteurized cheeses contaminated during the packaging process. After consumption, *Listeria* expresses virulence genes that facilitate invasion into epithelial cells. When high inocula are ingested, a self-limited febrile gastroenteritis can result, although this is seldom diagnosed, since specialized culture media are required to grow it in stool. The greater concern with listeriosis is the potential for disseminated disease, particularly in immunocompromised hosts and in pregnancy. In the former, bacteremia and meningitis are the major concerns, with an overall mortality as high as 25–28% (Arslan et al. 2015, Amaya-Villar et al. 2010). In the latter, the organism readily crosses the placenta leading to fetal infection that is highly lethal during the first or second trimesters (Lamont et al. 2011). Diagnosis in these cases is made by blood or CSF cultures, and treatment is typically with ampicillin +/– gentamicin.

- **Cholera.** *Vibrio cholerae* causes cholera, a devastating and frequently epidemic illness characterized by profuse, fulminant watery diarrhea that can lead to death from volume depletion within hours. While milder and even asymptomatic infections occur, these are frequently not recognized outside of epidemics. The organism colonizes oceanic copepods and easily contaminates drinking water supplies. It is readily grown in culture and can often be identified by direct microscopy of stool. ORS when feasible, or intravenous hydration when not, is the cornerstone of therapy. Tetracycline, azithromycin, and ciprofloxacin are options for antibiotic treatment, as they shorten the illness, although the disease is self-limited regardless if patients are appropriately volume repleted (Leibovici-Weissman et al. 2014).

- **Non-cholera Vibrios.** *Vibrio parahemolyticus* causes a self-limited inflammatory enterocolitis or watery diarrhea associated with uncooked seafood. The incidence has been increasing in the US (Newton et al. 2012). It is readily grown from stool on selective media. The benefit of antibiotic treatment has never been studied, but doxycycline or other antibiotics may be considered in severe cases. *V. parahemolyticus* can also cause extraintestinal infections, particularly cellulitis and sepsis. Another species, *V. vulnificus*, is more prone to such infections, particularly in patients with chronic liver disease or iron overload syndrome. Wound infections and severe cellulitis generally follow direct exposure to seawater, while sepsis typically follows ingestion of undercooked shellfish, and can have a rapid, fulminant course. For these reasons, people with chronic liver disease should always be advised to avoid eating raw or undercooked shellfish.

- **Yersinia.** *Y. enterocolitica* and *Y. pseudotuberculosis* frequently colonize livestock, particularly pork and consumption of undercooked pork is a major risk factor for their acquisition. Both species can cause a self-limited gastroenteritis that can be indistinguishable from other infectious diarrhea. Some features that are particularly characteristic of yersiniosis are a longer incubation period, more subacute onset of illness, and longer duration of illness than other enteric bacterial pathogens. Pharyngitis occurs in around 20% of cases, which is unusual in other diarrheal illnesses. Another characteristic syndrome is pseudoappendicitis due to inflamed lymphatic tissue; this is frequently identified at the time of surgery in patients with clinically suspected appendicitis. *Yersinia* grows in stool culture and can be routinely identified, particularly with cold enrichment and extended duration. Antibiotic treatment for yersiniosis has never been shown to be effective, although patients with bacteremia or sepsis should be treated.

- **Brucellosis.** *Brucella abortus* and other species cause illness in domestic cattle, goats, and other livestock, and consumption of unpasteurized milk or undercooked meat is the primary means of transmission. Brucellosis can be an acute febrile illness with septicemia, and can cause chronic focal infections, including endocarditis, osteomyelitis, and hepatitis. It grows well with modern blood culture techniques, but chronic cases often require serologic diagnosis. Treatment is generally with combination antibiotics, including aminoglycosides, rifampin, or doxycycline.

- **Clostridium difficile.** While the best recognized major cause of health care-associated diarrhea, *C. difficile* colonization is not uncommonly found in healthy people and in hospitalized patients at the time of admission, suggesting community acquisition. Since spores have frequently been identified in meat products (pork, beef, and poultry) it is believed that dietary acquisition is common (Varshney et al. 2014). However, perturbation of commensal intestinal bacteria, typically through antibiotics, is almost always required for disease to develop. The best protection against *C. difficile* is through careful infection control practices in health care settings and avoidance of unnecessary antibiotic use. *C. difficile* can be successfully treated in most cases with metronidazole, vancomycin, or fidaxomicin, although relapses are common.

Protozoa

- *Giardia lamblia.* Giardiasis is acquired after ingestion of cysts in contaminated food or water, or from direct person-to-person spread. Since it is endemic in wildlife mammalian species, contamination of groundwater is common. Infections occur sporadically and in outbreaks, as well as in travelers to developing areas. Filtration or cooking of water are the most effective ways to prevent transmission; *Giardia* is relatively resistant to chlorination. The most common manifestation is an acute diarrheal illness, frequently characterized by flatulence, bloating, and steatorrhea, believed to be due to small intestinal villus blunting leading to malabsorption of carbohydrates and lipids. In some cases, particularly with immune compromise, the illness is prolonged. Metronidazole for 7 days was historically recommended for giardiasis, but newer drugs have the advantage of shorter treatment duration and/or better side effect profile. These include tinidazole, nitazoxanide, and albendazole (Byrd-Leifer et al. 2001). Relapses can occur after successful metronidazole treatment, and drug resistance can develop. The optimal treatment in these cases remains unknown.

- *Cryptosporidium.* This protozoan is fairly ubiquitous in wildlife, although there is some degree of species specificity. The major species infecting humans is *C. hominis* (formerly *C. parvum* genotype 1). It can be found in domestic animals and wildlife and frequently contaminates groundwater. It is highly resistant to chlorination, leading to multiple documented outbreaks as well as sporadic disease. It can be removed by filtration (using a 1 μm filter) or inactivated by boiling. Cryptosporidiosis is usually a self limited diarrheal illness similar to giardiasis in immune competent children and adults, although symptoms can be prolonged in some cases. In contrast, infection in the setting of advanced HIV/AIDS or other severe immune defects can be devastating or frequently even lethal. Microscopy using a modified acid-fast stain was the historical gold standard for diagnosis, although PCR and commercial EIA assays are available with excellent sensitivity and specificity (Checkley et al. 2015, Van den Bossche et al. 2015). Treatment may be considered with nitazoxanide, which provides modest symptomatic benefit in immune competent patients (Cantley 2002). However, patients with HIV infection, particularly in those with CD4 counts less than 50/μl, do not respond well (DuPont et al. 1992). The mainstay of treatment for cryptosporidiosis in HIV/ AIDS is therefore antiretroviral therapy, which if successful can lead to lasting remission, although relapses occur if viral suppression is stopped (Carter et al. 2001).

- *Entamoeba histolytica.* This invasive protozoal pathogen is highly infectious and easily spreads person-to-person or through food or water contaminated by human feces. It can cause asymptomatic colonization, colitis (typically with dysentery), or invasive disease (predominantly liver abscess) with significant mortality. Historically it was diagnosed by microscopy, although a closely related nonpathogenic species, *Entamoeba dispar*, has an identical appearance. For this reason, many laboratories have moved to commercial EIA or PCR for diagnosis. While metronidazole is historically the recommended therapy, a meta-analysis of eight RCTs found that tinidazole is equally effective but with a shorter treatment duration and better side effect profile (Chan et al. 2001). A luminal agent such as paromomycin should be given to clear cysts after treatment of invasive disease.

- *Toxoplasma gondii. Toxoplasma* is an apicomplexan parasite with a highly promiscuous infectious ability. The definitive hosts are felines, including domestic and wild cats, who harbor the sexual forms and excrete infectious oocysts. Ingestion of these oocysts in contaminated food or drinking water can lead to human infection. In addition, wild and domestic livestock can become infected and form tissue cysts in muscle, which are also infectious when consumed. Consumption of undercooked meat is the major means of acquisition in the U.S. (Jones et al. 2009). The majority of infections are asymptomatic, although a minority of people develop a mild acute syndrome characterized by lymphadenopathy that persists for weeks or months. Rare severe organ involvement can occur and occasionally immune competent people develop acute or recurrent eye involvement (chorioretinitis). Infection in pregnancy can lead to significant fetal infection. Severely immunocompromised people are at risk of disseminated recurrent disease, which presents most commonly as a focal encephalitis. Toxoplasmosis is typically diagnosed by serology or characteristic findings on funduscopic examination or brain imaging in the appropriate clinical setting, but tissue identification remains the gold standard when required. Primary toxoplasmosis and ocular toxoplasmosis do not generally require treatment (Gilbert et al. 2002), but invasive, recurrent disease or illness in pregnancy should be treated; sulfonamides are the cornerstone of therapy and several effective combination regimens are available.

- *Other protozoa. Cyclospora cayetanensis* was first identified in the 1980s and caused several large foodborne outbreaks in the 1990s, leading to a greater understanding of its importance as a foodborne pathogen. It is endemic in a number of developing areas and is likely spread through direct person-to-person contact or contamination of food/

water with human feces rather than natural animal reservoirs. It causes a diarrheal illness with prominent fatigue, anorexia, and weight loss with an unusually prolonged duration (average more than three weeks) (Hoge et al. 1995). Diagnosis is by microscopy, using acid-fast stains, or fluorescence microscopy to detect autofluorescence. Treatment is generally recommended; TMP-SMX is the drug of choice, although ciprofloxacin is moderately effective for patients allergic to TMP-SMX. *Cystisospora belli* (formerly *Isospora*) causes a similar self-limited illness to *Cyclospora* in immunocompetent hosts and is also diagnosed by acid-fast staining and microscopy. In AIDS, illness can be prolonged or fatal like *Cryptosporidium*. Treatment is with TMP-SMX, and antiretroviral therapy is critical in the setting of AIDS to facilitate cure. *Trypanosoma cruzi* causes Chagas' disease, a chronic destruction of myenteric plexus in vital organs leading to cardiomyopathy and/or chronic intestinal disease (megacolon, mega-esophagus). It is ordinarily spread through the bite of a triatomine bug but can be acquired through ingestion of food contaminated with bug feces. *Blastocystis hominis* and *Dientamoeba fragilis* are protozoa of questionable clinical significance since they are frequently identified in asymptomatic individuals and treatment trials have been problematic (Stensvold et al. 2010).

Helminths. There are a number of important helminthic pathogens acquired primarily through ingestion of contaminated food and water. These include nematodes (roundworms), trematodes (flukes), and cestodes (tapeworms). Most are diagnosed through microscopy on formalin-fixed stool, although some have extraintestinal life cycles that require a different approach to diagnosis. The major pathogens in these categories are described below:

- Nematodes

  - *Ascaris lumbricoides*, or the giant roundworm, is acquired through ingestion of food (typically vegetables) or water contaminated with human feces. It is one of the most prevalent infections, with possibly over 1 billion people infected worldwide (Anonymous a). The worms live several years in the small intestine and most infestations are asymptomatic, but can cause illness when they migrate through narrow openings like the bile duct or pancreatic duct, or when there is a large enough worm burden to cause intestinal obstruction or chronic inflammation. Treatment is with mebendazole or albendazole.

  - *Trichuris trichiura*, or the whipworm, has a similar lifecycle and prevalence to *Ascaris* and is acquired through ingestion of food or water contaminated by human stool. Most cases are asymptomatic, but heavy infestations can lead to symptomatic colitis with increased stooling with blood or mucus and, in the worst cases, rectal prolapse. Treatment is with mebendazole or albendazole.

o *Trichinella* encysts in muscle tissue and is typically acquired through ingestion of undercooked meat. While pork was historically the major source, commercial swine in North America now have much less *Trichinella* and wild game (such as bear) are the most common source (Anonymous b). After ingestion, the worms encyst and penetrate the intestinal mucosa, which can cause a nonspecific mild gastroenteritis. The major symptoms develop later as larvae disseminate and enter muscle tissue, leading to muscle and/or joint pains, which can be severe and incapacitating in the worst cases. Periorbital edema and eosiniphilia are additional clues to infection. The diagnosis can be made by serology or demonstration of larvae on muscle biopsy. The symptoms are self-limited and resolve after muscle encystation is complete; treatment with albendazole plus corticosteroids is recommended in severe cases and those with visceral organ involvement (Dupouy-Camet et al. 2002).

o *Anisakis* is a common parasite in marine mammals, who shed viable eggs that enter the aquatic food chain as larvae and encyst in predator fish. Consumption of raw or undercooked fish allows the encysted larvae to emerge and attempt to penetrate the gastric or intestinal mucosa, leading to a florid inflammatory reaction as the larvae die. Gastric aniskiasis typically presents hours after consumption with severe epigastric pain, nausea, and vomiting, occasionally with expulsion of the worm. Intestinal disease takes longer to develop and can be more difficult to diagnose. Diagnosis is made by direct visualization of the worm either in vomitus or by endoscopy, and endoscopic removal is the treatment of choice. The illness can be prevented by adequate cooking or freezing of the fish prior to consumption, either of which can kill the larvae; the larval cysts are visible to the naked eye and careful inspection of fish filets can also reduce transmission. Capillariasis is a similar infection related to consumption of raw or undercooked freshwater fish and is endemic in the Philippines and parts of southeast Asia; it causes a severe, chronic diarrheal illness that can be fatal eventually if untreated.

o *Dracunculiasis* or Guinea worm is acquired through consumption of larval-infected copepods in fresh water. The larvae penetrate the intestinal mucosa and the female worms migrate to subcutaneous tissues, eventually producing a very painful local nodule through which the worm emerges to lay eggs. This produces a burning sensation relieved by immersion in cold water; live larvae are thus released into the water to complete the life cycle. Treatment is by surgical excision of the nodule, or by gradual extraction of the worm on a stick over a period of days. Infection can be completely prevented by provision of clean drinking water; even filtration of impure water sources is effective because it removes copepods. Introduction of these simple interventions is one of the great success stories of recent global health policy, leading to near-eradication of dracunculiasis with likelihood of complete eradication in the next few years.

- o *Angiostrongyliasis* is cased by ingestion of undercooked snails, slugs, or mollusks infected with the organism, or perhaps more commonly, vegetables on which these organisms have excreted larvae. This leads to a granulomatous eosinophilic enterocolitis that can be severe in the worst cases. Since eggs are not shed in stool, the diagnosis is by intestinal biopsy, but there is no effective treatment and the illness is self-limited.

- • Cestodes

- o *Taenia solium* (cysticercosis) is pathogenic in humans because of its propensity to cause tissue cysts, particularly in the brain, leading to seizures and other complications. The definitive host (meaning the animal that harbors the adult worm) is swine; humans are incidentally infected when they consume eggs in fecally-contaminated food or water, rather than from eating undercooked pork. These eggs excyst and larvae migrate throughout the body where they eventually die, leaving a walled off cyst that calcifies over time. While the living worm stage is occasionally symptomatic, depending on its location in the brain and the extent of inflammation, frequently all the worms are dead by the time seizures begin to develop and at that point there is little evidence of the benefit of the treatment (Abba et al. 2010), although some experts still recommend therapy with albendazole plus corticosteroids as it can reduce the total number of lesions (Baird et al. 2013). A presumptive diagnosis can be made on the basis of imaging plus serology on blood; tissue demonstration of the worms is the gold standard.

- o *Echinococcus* is a tapeworm whose definitive host is canines. The typical secondary hosts are either sheep or cervid mammals whose grazing involves ingestion of contaminated canine feces. People who farm infected sheep can themselves be infected when their dogs ingest contaminated meat during slaughter, and then expose their owners to feces. In the case of *E. gransulosus*, the worms migrate into tissue (most commonly liver) and over years (or decades) produce a slowly expanding tissue cyst that eventually becomes symptomatic when it is large enough to press on adjacent tissues. *E. multilocularis* produces a more aggressive disease with multiple cysts that expand more rapidly. Diagnosis can be made on the basis of serology plus imaging, although demonstration of scoleces on biopsy or aspiration of cyst fluid confirms the diagnosis. The optimal therapy is either total surgical resection or PAIR (Percutaneous Aspiration, Injection of scolecidal liquid, and Reaspiration); these are usually accompanied by albendazole or mebendazole in case cyst contents spill and spread infection. Pharmacological treatment on its own is curative in fewer than half of the cases (Kapan et al. 2008).

o *Human tapeworms* including *Taenia saginata* (beef tapeworm), *Diphyllobothrium latum* (fish tapeworm), and *Hymenolepsis nana* (dwarf tapeworm) are acquired from ingesting undercooked meat from animals that harbor infectious cysts in muscle. This results in growth of an adult tapeworm, that sheds eggs into stool, completing the life cycle when human feces enters the animal food chain. Infections are generally asymptomatic but can be disturbing when large sections of worms (or entire worms) are shed in stool. Long term infection can contribute to growth impairment or vitamin deficiencies and treatment with praziquantel or niclosamide is very effective.

- **Trematodes**

o *Liver flukes* are flatworms that infect freshwater fish, forming cysts in muscle; when raw or undercooked fish is eaten, the cysts rupture in the stomach, allowing flukes to migrate to the liver and biliary system, where the adults reside and lay eggs that are shed in stool. Diagnosis is made easily by identification of these eggs using microscopy. Infections with *Clonorchis* and *Opisthorchis* species are quite common in southeast Asia, particularly poorer areas where human waste is not treated properly. Chronic infection leads to biliary inflammation that can cause intermittent obstruction and eventually cholangiocarcinoma. Treatment with praziquantel is very effective. *Fasciola hepatica* is a similar fluke, although rather than proceeding through a fish intermediate host, humans and other animals ingest infections metacercariae in drinking water or on contaminated freshwater vegetation (e.g., watercress). Hence, fascioliasis is common worldwide. Infection can cause chronic biliary disease, although malignancy is rare. Diagnosis is by stool microscopy or serology. Treatment is with triclabendazole.

○ *Lung flukes* (*Paragonimus westermanii* and others) use freshwater crustaceans as an intermediate host, and people are infected by consumption of raw or undercooked crab, crayfish, or other crustaceans. Infection is most common in Asia. The flukes go through an intestinal phase but have a propensity to migrate to the lungs, causing cough, shortness of breath, and infiltrates on chest Xray that can mimic tuberculosis. Diagnosis is made by identification of eggs in sputum or stool, and treatment is carried out with praziquantel.

Conclusions

Foodborne infections are a major cause of human illness, but they do not need to be. Provision of basic sanitation (latrines, clean drinking water) to everyone would dramatically reduce the burden of illness worldwide. While there have been successes in this regard (Guinea worm being the most recent example), other elements of development including urbanization and commercial animal husbandry have increased the burden of some illnesses and introduced new ones such as STEC O104:H4, *Cyclospora*, and drug-resistant *Campylobacter*. Careful monitoring of livestock and meat products can be effective, as demonstrated by the reduction in EHEC cases associated with ground beef, but it is unlikely that the food chain will ever be completely free of pathogens, meaning that water treatment, proper cooking of food, and careful hygiene by food preparers are likely to remain the best ways to prevent infection for years to come.

References

a. Available from http://www.cdc.gov/parasites/ascariasis/.

b. Available from http://www.cdc.gov/parasites/trichinellosis/epi.html.

c. Available from http://www.cdc.gov/rotavirus/clinical.html.

d. Available from http://www.who.int/mediacentre/factsheets/fs180/en/.

Abba, K., S. Ramaratnam and L.N. Ranganathan. 2010. Anthelmintics for people with neurocysticercosis. The Cochrane Database of Systematic Reviews (3): CD000215. doi (3) (Mar 17): CD000215.

Al Mamun, A.A., A. Tominaga and M. Enomoto. 1996. Detection and characterization of the flagellar master operon in the four shigella subgroups. Journal of Bacteriology. 178 (13 Jul): 3722–6.

Amaya-Villar, R., E. García-Cabrera, E. Sulleiro-Igual, P. Fernández-Viladrich, D. Fontanals-Aymerich, P. Catalán-Alonso, C. Rodrigo-Gonzalo de Liria, A. Coloma-Conde, F. Grill-Díaz, A. Guerrero-Espejo, J. Pachón and G. Prats-Pastor. 2010. Three-year multicenter surveillance of community-acquired *Listeria monocytogenes* meningitis in adults. BMC Infectious Diseases. 10 (Nov 11): 324.

Appleton, H. 1987. Small round viruses: Classification and role in food-borne infections. Ciba Foundation Symposium. 128: 108–25.

Arslan, F., E. Meynet, M. Sunbul, O.R. Sipahi, B. Kurtaran, S. Kaya, A.C. Inkaya, P. Pagliano, G. Sengoz, A. Batirel, B. Kayaaslan, O. Yıldız, T. Güven, N. Türker, I. Midi, E. Parlak, S. Tosun, S. Erol, A. Inan, N. Oztoprak, I. Balkan, Y. Aksoy, B. Ceylan, M. Yılmaz and A. Mert. 2015. The clinical features, diagnosis, treatment, and prognosis of neuroinvasive listeriosis: A multinational study. European Journal of Clinical Microbiology & Infectious Diseases: Official Publication of the European Society of Clinical Microbiology 34(6) (Jun): 1213–21.

Atmar, R.L., A.R. Opekun, M.A. Gilger, M.K. Estes, S.E. Crawford, F.H. Neill, S. Ramani, H. Hill, J. Ferreira and D.Y. Graham. 2014. Determination of the 50% human infectious dose for norwalk virus. The Journal of Infectious Diseases. 209(7) (Apr 1): 1016–22.

Baird, R.A., S. Wiebe, J.R. Zunt, J.J. Halperin, G. Gronseth and K.L. Roos. 2013. Evidence-based guideline: Treatment of parenchymal neurocysticercosis: Report of the guideline development subcommittee of the american academy of neurology. Neurology 80(15) (Apr 9): 1424–9.

Balasegaram, S., A.L. Potter, D. Grynszpan, S. Barlow, R.H. Behrens, L. Lighton, L. Booth, L. Inamdar, K. Neal, K. Nye, J. Lawrence, J. Jones, I. Gray, D. Tolley, C. Lane, B. Adak, A. Cummins and S. Addiman; Typhiod and Paratyphoid Reference Group, Health Protection Agency, London, England. 2012. Guidelines for the public health management of typhoid and paratyphoid in England: Practice guidelines from the national Typhoid and Paratyphoid Reference Group. The Journal of Infection. 65(3) (Sep): 197–213.

Bhan, M.K., P. Raj, M.M. Levine, J.B. Kaper, N. Bhandari, R. Srivastava, R. Kumar and S. Sazawal. 1989. Enteroaggregative *Escherichia coli* associated with persistent diarrhea in a cohort of rural children in india. Journal of Infectious Diseases. 159(6 Jun): 1061–4.

Bhatnagar, S., M.K. Bhan, H. Sommerfelt, S. Sazawal, R. Kumar and S. Saini. 1993. Enteroaggregative *Escherichia coli* may be a new pathogen causing acute and persistent diarrhea. Scandinavian Journal of Infectious Diseases. 25(5): 579–83.

Bhatnagar, S., K.D. Singh, S. Sazawal, S.K. Saxena and M.K. Bhan. 1998. Efficacy of milk versus yogurt offered as part of a mixed diet in acute noncholera diarrhea among malnourished children. Journal of Pediatrics. 132(6): 999–1003.

Bian, Z.M., S.G. Elner, A. Yoshida and V.M. Elner. 2004. Differential involvement of phosphoinositide 3-kinase/Akt in human RPE MCP-1 and IL-8 expression. Invest. Ophthalmol. Vis. Sci. 45(6): 1887–96.

Birkenkamp, K.U., L.M. Tuyt, C. Lummen, A.T. Wierenga, W. Kruijer and E. Vellenga. 2000. The p38 MAP kinase inhibitor SB203580 enhances nuclear factor-kappa B transcriptional activity by a non-specific effect upon the ERK pathway. Br. J. Pharmacol. 131(1) (Sep Journal Article): 99–107.

Bouckenooghe, A., H. DuPont, Z. Jiang, J. Adachi, J. Mathewson, M. Verenkar, S. Rodrigues and R. Steffen. 1999. Markers of enteric inflammation in enteroaggregative *Escherichia coli* diarrhea in travelers. Paper Presented at 37th Annual Meeting of the Infectious Diseases Society of America.

Bouckenooghe, A.R., H.L. Dupont, Z.D. Jiang, J. Adachi, J.J. Mathewson, M.P. Verenkar, S. Rodrigues and R. Steffen. 2000. Markers of enteric inflammation in enteroaggregative *Escherichia coli* diarrhea in travelers. Am. J. Trop. Med. Hyg. 62(6) (Jun): 711–3.

Braegger, C.P., S. Nicholls, S.H. Murch, S. Stephens and T.T. MacDonald. 1992. Tumour necrosis factor alpha in stool as a marker of intestinal inflammation. Lancet 339 (8785 Jan 11): 89–91.

Brooks, S.A., J.E. Connolly and W.F. Rigby. 2004. The role of mRNA turnover in the regulation of tristetraprolin expression: Evidence for an extracellular signal-regulated kinase-specific, AU-rich element-dependent, autoregulatory pathway. J. Immunol. 172(12): 7263–71.

Brown, Z., R. Robson and J. Westwick. 1993. L-arginine/nitric oxide pathway: A possible signal transduction mechanism for the regulation of the chemokine IL-8 in human mesangial cells. pp. 65–75. *In:* I. Lindley (ed.). The Chemokines. Plenum Press, New York.

Burke, D. and A. Axon. 1987. Ulcerative colitis and *Escherichia coli* with adhesive properties. J. Clin. Path. 40: 782–6.

Burke, D., S. Clayden and A. Axon. 1988. Sulphasalazine does not select for *Escherichia coli* with adhesive properties in ulcerative colitis. Lancet. 2(1867): 966.

Byrd-Leifer, C.A., E.F. Block, K. Takeda, S. Akira and A. Ding. 2001. The role of MyD88 and TLR4 in the LPS-mimetic activity of taxol. Eur. J. Immunol. 31(8) (Aug): 2448–57.

Cantley, L.C. 2002. The phosphoinositide 3-kinase pathway. Science. 296(5573): 1655–7.

Carter, A.B., L.A. Tephly and G.W. Hunninghake. 2001. The absence of activator protein 1-dependent gene expression in THP-1 macrophages stimulated with phorbol esters is due to lack of p38 mitogen-activated protein kinase activation. J. Biol. Chem. 276(36): 33826–32.

Chan, F.K., R.M. Siegel, D. Zacharias, R. Swofford, K.L. Holmes, R.Y. Tsien and M.J. Lenardo. 2001. Fluorescence resonance energy transfer analysis of cell surface receptor interactions and signaling using spectral variants of the green fluorescent protein. Cytometry. 44(4) (Aug 1): 361–8.

Chan, K.N., A.D. Phillips, S. Knutton, H.R. Smith and J.A. Walker-Smith. 1994. Enteroaggregative *Escherichia coli*: Another cause of acute and chronic diarrhoea in england? Journal of Pediatric Gastroenterology & Nutrition. 18 (1 Jan): 87–91.

Checkley, W., R. Gilman, L. Epstein, F. Diaz, L. Cabrera, R. Black and C. Sterling. 1996. The adverse effect of cryptosporidium parvum infection on the growth of children. Paper presented at Fifth Annual Meeting of the International Centers for Tropical Diseases Research.

Checkley, W., A.C. White, Jr., D. Jaganath, M.J. Arrowood, R.M. Chalmers, X.M. Chen, R. Fayer, J.K. Griffiths, R.L. Guerrant, L. Hedstrom, C.D. Huston, K.L. Kotloff, G. Kang, J.R. Mead, M. Miller, W.A. Petri, Jr., J.W. Priest, D.S. Roos, B. Striepen, R.C. Thompson, H.D. Ward, W.A. Van Voorhis, L. Xiao, G. Zhu and E.R. Houpt. 2015. A review of the global burden, novel diagnostics, therapeutics, and vaccine targets for cryptosporidium. The Lancet. Infectious Diseases. 15(1) (Jan): 85–94.

Chiao, D.J., J.J. Wey, P.Y. Tsui, F.G. Lin and R.H. Shyu. 2013. Comparison of LFA with PCR and RPLA in detecting SEB from isolated clinical strains of *Staphylococcus aureus* and its application in food samples. Food Chemistry. 141(3) (Dec 1): 1789–95.

Choi, E.K., H.J. Park, J.S. Ma, H.C. Lee, H.C. Kang, B.G. Kim and I.C. Kang. 2004. LY294002 inhibits monocyte chemoattractant protein-1 expression through a phosphatidylinositol 3-kinase-independent mechanism. FEBS Lett. 559(1-3) (Feb 13): 141–4.

Cooke, E. 1968. Properties of strains of *Escherichia coli* isolated from the faeces of patients with ulcerative colitis, patients with acute diarrhoea, and normal persons. J. Path. Bact. 95: 102–13.

Coopersmith, C.M., P.E. Stromberg, W.M. Dunne, C.G. Davis, D.M. Amiot 2nd, T.G. Buchman, I.E. Karl and R.S. Hotchkiss. 2002. Inhibition of intestinal epithelial apoptosis and survival in a murine model of pneumonia-induced sepsis. Jama. 287(13) (Apr 3): 1716–21.

Deng, W., J.L. Puente, S. Gruenheid, Y. Li, B.A. Vallance, A. Vázquez, J. Barba, J.A. Ibarra, P. O'Donnell, P. Metalnikov, K. Ashman, S. Lee, D. Goode, T. Pawson and B.B. Finlay. 2004. Dissecting virulence: Systematic and functional analyses of a pathogenicity island. Proc. Natl. Acad. Sci. USA. 101(10) (Mar 9 Journal Article): 3597–602.

Dunne, A., M. Ejdeback, P.L. Ludidi, L.A. O'Neill and N.J. Gay. 2003. Structural complementarity of Toll/interleukin-1 receptor domains in toll-like receptors and the adaptors mal and MyD88. Journal of Biological Chemistry. 278(42): 41443–51.

Dunne, A. and L.A. O'Neill. 2003. The interleukin-1 receptor/Toll-like receptor superfamily: Signal transduction during inflammation and host defense. Sci. STKE. 2003(171): re3.

DuPont, H.L., C.D. Ericsson, M.J. Farthing, S. Gorbach, L.K. Pickering, L. Rombo, R. Steffen and T. Weinke. 2009. Expert review of the evidence base for self-therapy of travelers' diarrhea. Journal of Travel Medicine. 16(3) (May–Jun): 161–71.

DuPont, H.L., C.D. Ericsson, J.J. Mathewson and M.W. DuPont. 1992. Five versus three days of ofloxacin therapy for traveler's diarrhea: A placebo-controlled study. Antimicrob. Agents Chemother. 36(1) (Jan): 87–91.

Dupouy-Camet, J., W. Kociecka, F. Bruschi, F. Bolas-Fernandez and E. Pozio. 2002. Opinion on the diagnosis and treatment of human trichinellosis. Expert Opinion on Pharmacotherapy. 3(8) (Aug): 1117–30.

Dutta, S., D. Sur, B. Manna, B. Sen, A.K. Deb, J.L. Deen, J. Wain et al. 2006. Evaluation of new-generation serologic tests for the diagnosis of typhoid fever: Data from a community-based surveillance in calcutta, india. Diagnostic Microbiology & Infectious Disease. 56(4) (Dec): 359–65.

Eaves-Pyles, T., K. Murthy, L. Liaudet, L. Virag, G. Ross, F.G. Soriano, C. Szabo and A.L. Salzman. 2001a. Flagellin, a novel mediator of salmonella-induced epithelial activation and systemic inflammation: I kappa B alpha degradation, induction of nitric oxide synthase, induction of proinflammatory mediators, and cardiovascular dysfunction. J Immunol. 166(2) (Jan 15): 1248–60.

Eaves-Pyles, T.D., H.R. Wong, K. Odoms and R.B. Pyles. 2001b. Salmonella flagellin-dependent proinflammatory responses are localized to the conserved amino and carboxyl regions of the protein. J. Immunol. 167(12) (Dec 15): 7009–16.

Fisman, D.N., A.L. Greer, G. Brouhanski and S.J. Drews. 2009. Of gastro and the gold standard: Evaluation and policy implications of norovirus test performance for outbreak detection. Journal of Translational Medicine. 7(Mar 26): 23,5876-7-23.

Freedman, S.B. 2007. Acute infectious pediatric gastroenteritis: Beyond oral rehydration therapy. Expert Opinion on Pharmacotherapy. 8(11) (Aug): 1651–65.

From: The Indian Society of Critical Care Medicine Tropical fever Group, S. Singhi, D. Chaudhary, G.M. Varghese, A. Bhalla, N. Karthi, S. Kalantri et al. 2014. Tropical fevers: Management guidelines. Indian Journal of Critical Care Medicine : Peer-Reviewed, Official Publication of Indian Society of Critical Care Medicine. 18(2) (Feb): 62–9.

Gabbay, Y.B., A.C. Linhares, E.L. Cavalcante-Pepino, L.S. Nakamura, D.S. Oliveira, L.D. da Silva, J.D. Mascarenhas, C.S. Oliveira, T.A. Monteiro and J.P. Leite. 2007. Prevalence of human astrovirus genotypes associated with acute gastroenteritis among children in belem, brazil. Journal of Medical Virology. 79(5) (May): 530–8.

Gilbert, R.E., S.E. See, L.V. Jones and M.S. Stanford. 2002. Antibiotics versus control for toxoplasma retinochoroiditis. The Cochrane Database of Systematic Reviews. (1)(1): CD002218.

Guarino, A., S. Ashkenazi, D. Gendrel, A. Lo Vecchio, R. Shamir and H. Szajewska. 2014. European society for pediatric gastroenterology, hepatology, and Nutrition/European society for pediatric infectious diseases evidence-based guidelines for the management of acute gastroenteritis in children in europe: Update 2014. Journal of Pediatric Gastroenterology and Nutrition. 59(1) (Jul): 132–52.

Guerrant, R.L., M.D. DeBoer, S.R. Moore, R.J. Scharf and A.A. Lima. 2013. The impoverished gut—a triple burden of diarrhoea, stunting and chronic disease. Nature Reviews. Gastroenterology & Hepatology. 10(4) (Apr): 220–9.

Guerrant, R.L., T. Van Gilder, T.S. Steiner, N.M. Thielman, L. Slutsker, R.V. Tauxe, T. Hennessy et al. 2001. Practice guidelines for the management of infectious diarrhea. Clinical Infectious Diseases. 32(3): 331–51.

Gurwith, M., W. Wenman, D. Hinde, S. Feltham and H. Greenberg. 1981. A prospective study of rotavirus infection in infants and young children. The Journal of Infectious Diseases. 144(3) (Sep): 218–24.

Hall, A.J., B.A. Lopman, D.C. Payne, M.M. Patel, P.A. Gastanaduy, J. Vinje and U.D. Parashar. 2013. Norovirus disease in the united states. Emerging Infectious Diseases. 19(8) (Aug): 1198–205.

Hoge, C.W., D.R. Shlim, M. Ghimire, J.G. Rabold, P. Pandey, A. Walch, R. Rajah, P. Gaudio and P. Echeverria. 1995. Placebo-controlled trial of co-trimoxazole for cyclospora infections among travellers and foreign residents in nepal. Lancet. 345(8951) (Mar 18): 691–3.

Jones, J.L., V. Dargelas, J. Roberts, C. Press, J.S. Remington and J.G. Montoya. 2009. Risk factors for toxoplasma gondii infection in the united states. Clinical Infectious Diseases: An Official Publication of the Infectious Diseases Society of America. 49(6) (Sep 15): 878–84.

Kapan, S., A.N. Turhan, M.U. Kalayci, H. Alis and E. Aygun. 2008. Albendazole is not effective for primary treatment of hepatic hydatid cysts. Journal of Gastrointestinal Surgery: Official Journal of the Society for Surgery of the Alimentary Tract. 12(5) (May): 867–71.

Krajden, M., M. Brown, A. Petrasek and P.J. Middleton. 1990. Clinical features of adenovirus enteritis: A review of 127 cases. The Pediatric Infectious Disease Journal. 9(9) (Sep): 636–41.

Lamont, R.F., J. Sobel, S. Mazaki-Tovi, J.P. Kusanovic, E. Vaisbuch, S.K. Kim, N. Uldbjerg and R. Romero. 2011. Listeriosis in human pregnancy: A systematic review. Journal of Perinatal Medicine. 39(3) (May): 227–36.

Leibovici-Weissman, Y., A. Neuberger, R. Bitterman, D. Sinclair, M.A. Salam and M. Paul. 2014. Antimicrobial drugs for treating cholera. The Cochrane Database of Systematic Reviews. 6 (Jun 19): CD008625.

Morgan, O., L. Milne, S. Kumar, D. Murray, W. Man, M. Georgiou, N.Q. Verlander, E. de Pinna, and M. McEvoy. 2007. Outbreak of salmonella enteritidis phage type 13a: Case-control investigation in hertsmere, united kingdom. Euro Surveillance: Bulletin Europeen Sur Les Maladies Transmissibles = European Communicable Disease Bulletin. 12(7) (Jul): E9–10.

Nataro, J.P. and T.S. Steiner. 2002. Enteroaggregative and diffusely adherent *Escherichia coli*. pp. 189–207. *In*: M.S. Donnenberg (ed.). *Escherichia coli*: Virulence Mechanisms of a Versatile Pathogen. San Diego: Elsevier Science (USA).

Newton, A., M. Kendall, D.J. Vugia, O.L. Henao and B.E. Mahon. 2012. Increasing rates of vibriosis in the united states, 1996–2010: Review of surveillance data from 2 systems. Clinical Infectious Diseases: An Official Publication of the Infectious Diseases Society of America. 54 Suppl. 5(Jun): S391–5.

Porter, C.K., N. Thura, R.T. Ranallo and M.S. Riddle. 2013. The shigella human challenge model. Epidemiology and Infection. 141(2) (Feb): 223–32.

Safdar, Nasia, Adnan Said, Ronald E. Gangnon and Dennis G. Maki. 2002. Risk of hemolytic uremic syndrome after antibiotic treatment of *Escherichia coli* O157:H7 enteritis: A meta-analysis. Jama. 288(8) (Aug 28): 996–1001.

Sahin, O., T.Y. Morishita and Q. Zhang. 2002. Campylobacter colonization in poultry: Sources of infection and modes of transmission. Animal Health Research Reviews/Conference of Research Workers in Animal Diseases. 3(2) (Dec): 95–105.

Spinale, J.M., R.L. Ruebner, L. Copelovitch and B.S. Kaplan. 2013. Long-term outcomes of shiga toxin hemolytic uremic syndrome. Pediatric Nephrology (Berlin, Germany). 28(11) (Nov): 2097–105.

Steiner, T.S., A.A. Lima, J.P. Nataro and R.L. Guerrant. 1998. Enteroaggregative *Escherichia coli* produce intestinal inflammation and growth impairment and cause interleukin-8 release from intestinal epithelial cells. J. Infect. Dis. 177(1): 88–96.

Stensvold, C.R., H.V. Smith, R. Nagel, K.E. Olsen and R.J. Traub. 2010. Eradication of blastocystis carriage with antimicrobials: Reality or delusion? Journal of Clinical Gastroenterology. 44(2) (Feb): 85–90.

Van den Bossche, D., L. Cnops, J. Verschueren and M. Van Esbroeck. 2015. Comparison of four rapid diagnostic tests, ELISA, microscopy and PCR for the detection of giardia lamblia, *cryptosporidium* spp. and *Entamoeba histolytica* in feces. Journal of Microbiological Methods. 110(Mar): 78–84.

Varshney, J.B., K.J. Very, J.L. Williams, J.P. Hegarty, D.B. Stewart, J. Lumadue, K. Venkitanarayanan and B.M. Jayarao. 2014. Characterization of *Clostridium difficile* isolates from human fecal samples and retail meat from pennsylvania. Foodborne Pathogens and Disease. 11(10) (Oct): 822–9.

White, P.L., A.L. Naugle, C.R. Jackson, P.J. Fedorka-Cray, B.E. Rose, K.M. Pritchard, P. Levine, P.K. Saini, C.M. Schroeder, M.S. Dreyfuss, R. Tan, K.G. Holt, J. Harman and S. Buchanan. 2007. Salmonella Enteritidis in meat, poultry, and pasteurized egg products regulated by the U.S. Food Safety and Inspection Service, 1998 through 2003. Journal of Food Protection. 70(3) (Mar): 582–91.

CHAPTER 8

Antibiotic Use in Animal Feed and its Impact on Antibiotic Resistance in Human Pathogens

Daniel P. Ballard, Emily A. Peterson, Joseph L. Nadler
and *Nancy M. Khardori**

Introduction

Along with the development of vaccines, antimicrobial agents are widely regarded as one of the most important advances in modern medicine. Prior to the discovery of antibiotics, common infections often resulted in death. Antibiotics have made possible the treatment and prevention of diseases which were previously responsible for incredible morbidity and mortality both in children and adults. After Alexander Fleming discovered penicillin in 1928, the drug was celebrated as a miracle cure. Today, antimicrobial drugs have become easily available, relatively non-toxic and have therefore fallen prey to overuse and abuse. The collateral damage associated with their appropriate use, overuse and abuse are all associated with the development of resistance to these lifesaving agents.

Division of Infectious Diseases, Department of Internal Medicine Eastern Virginia, Medical School, Norfolk, Virginia.
* Corresponding author: khardoNM@evms.edu

Soon after penicillin use became widespread in the 1940s, the problem of antibiotic resistance became apparent. In his Nobel lecture, Fleming proved his incredible foresight in the light of today's battle with antimicrobial resistance:

"There may be a danger, though, in underdosage. It is not difficult to make microbes resistant to penicillin in the laboratory by exposing them to concentrations not sufficient to kill them, and the same thing has occasionally happened in the body. The time may come when penicillin can be bought by anyone in the shops. Then there is the danger that the ignorant man may easily underdose himself and by exposing his microbes to non-lethal quantities of the drug make them resistant" (Fleming 1945).

Despite attempts to spread awareness of the problem throughout the medical field, irresponsible prescribing remains a large contributor to the development of antimicrobial resistance. Multi-drug resistant organisms are now commonplace in all types of healthcare facilities around the world, leading to treatment failures, increased severity of disease and fatal outcomes. The increase in global travel and trade makes it possible for this resistance to move among countries and continents.

But while a number of national initiatives have been developed to address inappropriate use of antibiotics for human disease, the overuse of antibiotics in animal food production remains a largely unaddressed contributor to the development and spread of antibiotic-resistant organisms. In 2011, nearly four times as many antibiotics were sold for use in livestock and poultry compared to those sold for treating human illness (PEW 2014). Coupled with the fact that animals used for food production vastly outnumber humans in the United States, the enormity of the problem becomes clear (Humane Society 2014, Turnidge 2004). Since the approval of antimicrobial drugs for use as growth promoters in the 1950s, numerous U.S. and international organizations have struggled to raise awareness about the importance of this practice in the development of antibiotic resistance.

The central tenets of the problem have been outlined in a number of reports by the FDA, the WHO, and other U.S. and international organizations. First, there is clear evidence that non-human usage of antimicrobial agents results in the development of drug-resistant organisms. Second, resistant bacteria and resistance genes are transmitted through the foodborne route from animal population to humans. Finally, these resistant pathogens cause adverse human health consequences, presenting a gloomy prospect for the future (WHO 2003).

But while strides have been made to curb the contribution of human medical care to the development of antibiotic resistance, the role of non-human use of antimicrobial drugs continues largely to fly under the radar. Antimicrobial drugs have been approved for use as growth promoters in livestock since the 1950s. Scientists have raised concern about the development of resistance through this practice since the 1960s. And yet, after 40 years of warnings and recommendations to reduce the use of antibiotics for growth promotion, the FDA has made little headway in this endeavor.

Over the last five years, there has been cause for optimism. In 2013, a bill was introduced to Congress that proposed increased oversight and regulation of antibiotic use for food-producing animals was introduced to the congress and importantly, placed the burden of proof of safety on drug manufacturers (House of Representatives 2013). Later the same year, the FDA released its strongest guidelines to date, which recommended voluntary changes to antibiotic labels and phasing out of medically important antimicrobial agents from use as growth promoters (FDA 2013). Although these efforts are unlikely to have a significant effect on the industry's consumption of antibiotics, they signal the movement of the conversation into the national spotlight. Using lessons learned from European legislative efforts, the momentum is gathering behind the movement to end the dangerous practice of widespread use of antibiotics in animal food production.

Antibiotic Use in the Feed and Drinking Water of Food-Producing Animals

The development of antimicrobial resistance among both human and animal pathogens represents a tremendous threat to global health. Antimicrobial use—whether for treatment, prophylaxis, or growth promotion—drives the selection of resistance genes in bacteria. While there is minimal data published regarding the quantity of antibiotics used to treat human infections, reports on the amount of antibiotics used in the U.S. for farm animals range from 17.8 to 24.6 million pounds annually. The overwhelming majority of these drugs are used for production purposes, including growth promotion and feed efficiency (Pew 2008). Estimates of human use, for comparison, range from 7.6 to 10.5 million pounds per year (FDA 2014, Oliver et al. 2011, Spellberg et al. 2013).

Inappropriate use of antibiotics to treat human diseases is a well-recognized source of antimicrobial resistance, and there are a number of major initiatives to promote the thoughtful use of antibiotics in clinical medicine. These efforts have already resulted in improved recognition and treatment of diseases for which antimicrobial drugs are not indicated

(Mainous et al. 2003). Most healthcare facilities have put antibiotic stewardship programs in place to drive appropriate and optimal use of antibiotics in the hospitalized patient population. However, the overuse and abuse of antibiotics in the outpatient settings remains largely unaddressed.

While the judicious use of antimicrobial drugs for the treatment of human disease remains an important part of the effort to slow the development of multi-drug resistant organisms, the impact of antimicrobial use in food production must not be underestimated. And unlike the complex prospect of changing clinical prescribing habits, reducing antibiotic use in industrial animal farming involves a simpler solution and carries little risk of unintended harm to human health.

In 1946, scientists at the University of Wisconsin reported their discovery that small quantities of antibiotics stimulate the growth of young animals (Moore et al. 1946). In the ten years following this publication, animal farms in developed countries worldwide adopted the practice of adding antibiotics to animal feed to promote growth (Smith 1968). The use of antibiotics for this purpose is now ubiquitous, and a number of studies have since supported this phenomenon of antibiotic-enhanced growth. Experiments have shown that low-dose antibiotic supplements allow for less feed with no decrease in growth, as well as improved quality and higher protein content in the meat of antibiotic-fed animals (Hughes and Heritage 2002).

The physiology underlying these effects remains unclear, but a number of theories have been suggested. Most experts agree that antibiotics most likely provide some form of protection from the unhygienic environments present on industrial animal farms. Some postulate that antibiotics may decrease the burden of microorganisms in animal feed, minimizing the nutrient loss from the feed. Alternatively, by eliminating pathogenic bacteria, they might reduce the catabolic effects of chronic inflammatory responses, which cause weight loss and muscle wasting (Chattopadhyay 2014).

But while the benefits of antimicrobial use for growth promotion may be appealing, the risks to human health, now and in the future, are far greater. Of the millions of pounds of antibiotics used annually to augment growth in farm animals, many are drugs that serve a critical role in human medicine (Mellon et al. 2001). In 2011, medically important antimicrobials accounted for 61% of all antimicrobials used in food-producing animals. Of these tetracyclines (68%) and penicillins (11%) were most common, followed by macrolides (7%) and sulfonamides (FDA 2014). Antimicrobials not currently used in humans way also select genes that lead to cross-resistance to those that are used to human infections.

The risk of this practice is two-fold. First, while the development of antimicrobial resistance among bacterial populations is a natural phenomenon, the rate at which resistance is manifested is directly related to the amount of selective pressure applied to the microbial population. Following this premise, the administration of antimicrobial agents to an entire flock or herd applies a significantly higher stress to the bacterial population than the use of these drugs to treat individuals or groups of animals.

The second, and perhaps greater contributor to the development of resistance is the duration of exposure and concentration of the antimicrobial used for growth promotion. Whereas the treatment or control of an illness is accomplished with a short-term, high dose antibiotic regimen, antimicrobial drugs are given in relatively low concentrations for prolonged periods of time when used for promotion of growth or food production efficiency (Moore et al. 1946, Gustafson and Bowen 1997). This low-level, sustained exposure to subinhibitory concentrations may produce a greater selective pressure, resulting in more numerous and diverse resistant organisms.

A number of studies have demonstrated (Alali et al. 2008) the ability for antimicrobial resistance to transfer between animal and human populations. Resistance among serotypes of *Salmonella, Campylobacter, Enterococci,* and *Escherichia coli* has been correlated to the use of broad-spectrum antimicrobials in livestock (WHO 1997). The development of multi-drug resistant (MDR) human pathogens results in adverse consequences including increased treatment failures and increased severity of disease. With the incidence of human infections caused by MDR pathogens increasing, and little progress in the development of new antibiotics, the number of viable therapeutic options is rapidly declining. Above all, even with the development of newer therapeutic agents, the antimicrobial resistance will continue to grow unless measures to prevent its occurrence in humans as well as animals are taken seriously. In a philosophical sense, discovering more antibiotics may result in adding fuel to the fire.

Antibiotic Use in Animal Food Production Results in the Development of Antimicrobial Resistance in Human Pathogens

Since the use of antibiotics for growth promotion became commonplace on industrial animal farms, scientists have raised concern that the practice results in the development of antimicrobial resistance. A number of studies conducted over the past 60 years have sought to demonstrate this relationship, and yet the quantity of antibiotics used for this purpose has grown exponentially (FDA 2014, Oliver et al. 2011).

One of the earliest studies in the United States to investigate the association between antimicrobial resistance and food production was conducted in 1976 by Dr. Stuart Levy of Tufts Medical School (Levy et al. 1976). Chickens on a rural (and previously antibiotic free) farm in Iowa were given feed supplemented with oxytetracycline. The chickens began to excrete tetracycline-resistant *E. coli* in their feces within 1–2 days. By three months, the chickens were excreting *E. coli* resistant to multiple other antibiotics, including sulfonamides, ampicillin, streptomycin, and carbenicillin. Perhaps the most alarming outcome of the study, farm personnel began to show increases in *E. coli* resistant to multiple antibiotics in their feces.

Since the publication of this landmark study, Levy has continued to champion the movement to reduce the amount of antimicrobial agents used for industrial animal farming. In a testimony given to the Subcommittee on Health of the U.S. House Committee on Energy and Commerce on July 14th of 2010, Levy established three areas of concern in relation to antimicrobial use in livestock and the impact on human health: (1) antibiotics are "societal drugs", and their use and effects—i.e., development of resistance—in one individual affects other individuals in the environment; (2) antibiotics are ecological agents that can effect change in populations in the environment; (3) the total amount of antibiotic used is not a good indicator of the changes to the environment; the distribution of the antibiotic will predict outcomes more reliably. Lower doses that do not kill bacteria contribute more significantly to development of resistance.

In 1984, in cooperation with the Seattle-King County Department of Public Health, the CDC published another landmark study. The study sought to determine the relationship between the occurrence of *Salmonella* and *Campylobacter jejuni* in foods of animal origin and the occurrence of human illness caused by these two organisms. Retail meats were sampled over a 20-month period with simultaneous investigation of *Salmonella* and *C. jejuni* enteritis cases in humans. The investigation found that *C. jejuni* was a more common cause of enteritis than *Salmonella*, and "does appear to flow from chickens to man via the consumption of poultry products (Seattle-King County Department of Public Health 1984)."

Now central to the argument against the use of antibiotics in animal feed, this study provided the initial evidence that bacteria can be transmitted from animals to humans via the food chain.

A number of studies have since supported these findings and have sought to make the further connection that antibiotic resistant bacteria lead to detrimental effects on human health (Helms et al. 2002, Martin et al. 2004, Nelson et al. 2004). In an investigation done in 1984 by the CDC, the fatality rate from infection due to antimicrobial-resistant *Salmonella* was 4.2% as opposed to 0.2% in non-microbial resistant bacteria. In addition it was found that food animals were the source of 69% of the antimicrobial-resistant infections as opposed to 46% of the susceptible outbreak strains (Holmberg et al. 1984).

Selection for and persistence of resistant bacteria in food-producing animals is further enhanced by subtherapeutic/subinhibitory doses of antimicrobial agents. Resistant organisms can then be transmitted via the food chain to humans and ultimately afflict them with the disease. Comparison of infections due to antimicrobial-resistant bacteria and antimicrobial-susceptible strains of the same bacteria shows that an infection with resistant bacteria leads to higher mortality, prolonged symptoms, and more hospitalizations. Patients who acquire these resistant infections have overall poorer outcomes. The clear cut adverse health and economic outcomes of acquiring such infections cannot be explained entirely by their tendency to infect the old and infirm (Holmberg et al. 1987). Because of the marked increases in adverse outcomes associated with antimicrobial-resistant bacteria, stemming the development of antimicrobial resistance and the routes by which pathogens develop this resistance is a topic of clear public health importance.

In addition to targeted studies that examine antimicrobial resistance as a result of use in animal production, there is evidence of increases in antimicrobial resistant infections in the United States as a whole. Between 1979–1980 it was found that 17% of *Salmonella* infections were antimicrobial resistant, compared to 31% from 1989–1990. Tetracycline resistance characterized the first half of the decade, and was followed by the development of resistance to more clinically important agents, such as ampicillin and gentamicin (Lee et al. 1994).

There is also evidence that antibiotic resistant microorganisms increased both in abundance and diversity in the microbiome of pig gut. There are highly variable regions within the 16S rRNA sequences that can be exploited with PCR for analysis of phylogenetic and metagenomic differences in the bacterial sequences. The increase found in abundance and diversity of the resistance genes was even measured above the high background level conferred from 50 years of using antimicrobials for their growth enhancing properties. Even more concerning was the increase in 'enriched genes' such as aminoglycoside O-phosphotransferases that conferred resistance to antibiotics not administered in the study. This demonstrates a potential for indirect selection of resistance to classes of antibiotics not even fed (Looft et al. 2011). These enriched sequences have been shown to export chemicals that may include antibiotics, allowing even bacteria that lack resistance genes to survive the pressure of antibiotics. This multidrug resistance is medically alarming for reasons already outlined.

While clearly the elimination or at least more careful consideration of the use of antibiotics in livestock feed solely for growth purposes is in order, even this practice does not stem the problem entirely. In a study conducted, comparing organically raised hogs with wild swine, it was found that although the organically raised hogs had not been exposed to antibiotics in their feed for over four years, each generation of new swine still had resistant strains of microbiota in their gut. This is due in part to the transmission of the gut microbiome from sow to piglet. In addition, the gut microbiota may still exchange resistance genes through plasmid recombination even in the absence of the selective pressure of antibiotics (Stanton et al. 2011).

It is also possible for strains of bacteria to pass from humans to animals and back again. In 2012 genetic fingerprinting was used to trace the origin of a strain of methicillin-resistant Staphylococcus aureus (MRSA) that was spreading in livestock and animals alike. In was found that humans originally passed a drug susceptible strain of *S. aureus* to animals and once the livestock acquired this strain the bacteria developed resistance to the antibiotics that were being fed to the livestock subtherapeutically. This strain of bacteria was then passed back to humans as MRSA. It is chains of transmission such as this that demonstrate the incredible adverse consequences of regular administration of antibiotics to livestock. Additionally, resistance genes can congregate on mobile genetic elements transferred between bacterium with relative ease, resulting in the rapid spread of resistance among bacterium living in the constant presence of antibiotics (Looft et al. 2011).

While there has been overwhelming evidence demonstrating the development of resistant bacteria in livestock and the transfer of this to humans, there are of course other methods of developing antimicrobial resistant bacteria. There are postulations that while antimicrobial resistance development in livestock has been proven to transfer to humans through the food chain, this does not often happen in nature. In addition much of antibiotic resistance development has arisen from human use and the development of infections in hospitals as well as nosocomial infections. Despite this, the troubling facts remain that antibiotic use is rampant in the livestock industry, it does contribute to antimicrobial resistance in animals, this resistance can be transferred to humans and this part of the chain is easily corrected.

The European Ban on Antibiotics in Food Animal Production

European states began the use of antibiotic growth promoters (AGPs) in the 1950s and 1960s with each state approving its own regulations for the use of antibiotics in animal feed (Castanon 2007). AGPs were shown to increase the rate of animal growth, reduce rates of diseases, and improve feed conversion. Soon after the approval of use, however, concerns began to be raised regarding the development of antibiotic resistance in human pathogens due to the use of AGPs. In 1969, the discovery of transferable oxytetracycline resistance in *Salmonella enterica* isolated from food-producing animals led to the creation of the Swann Committee by the British government. After investigation into the issue, the Swann Committee released a report calling for the separation of the usage of antimicrobials into two categories. Low-dose, non-therapeutic, non-prescribed usage as a feed additive and high-dose, therapeutic, prescribed usage. The report called for restricted use of AGPs in the non-therapeutic category to reduce the risk of bacteria developing resistance to drugs for human use. This report led to the removal of penicillin, tetracycline, and streptomycin from the list of approved AGPs in many European countries in 1972–1974 (Cogliani et al. 2011).

Other studies were conducted to test antibiotic resistance stemming from the use of AGPs. A 1975 study conducted by Stuart Levy at Tufts University found that low dose usage of oxytetracycline in chickens leads to multidrug resistance in *E. coli* that is quickly transferred to human farm dwellers not taking any antibiotics (Levy 1976). A much larger study conducted by Ruth Hummel, Helmut Tschäpe, and Wolfgang Witte at the Robert Koch Institute in Germany found the foremost selective force for streptogramin resistance genes in *Enterococcus faecium* in humans was the use of virginiamycin as an AGP in livestock (Cogliani et al. 2011).

In 1980, Sweden began to collect data regarding antimicrobial use in agriculture. In 1984, a report issued regarding consumer confidence in meat safety noted a drop after the public learned that roughly thirty tons of antimicrobial agents were being used for livestock annually. In 1986, Sweden became the first country to withdraw the use of AGPs from standard practice of food animal production. This was in part due to request from farmers after learning of the 1984 report.

At the time of the 1986 ban, Sweden did not have a resistance monitoring system in place; however, there was a system for monitoring antibiotic usage and services designed to educate farmers about responsible practices. These educational programs were directed to prevent infection, keep animals healthy and included guidelines on hygiene, medications, and feed. These efforts eventually led to a significant decrease in the usage of antibiotics. In 2000 Sweden established SVARM to monitor antimicrobial resistance in farm animals.

Through these measures, average sales of antibiotics for animals declined from forty-five tons to fifteen tons in 2009. However, this ban did not come without early issues, including *Clostridium perfringens*-associated diarrhea and necrotizing enteritis in poultry, weaning diarrhea in piglets and dysentery in pigs to be slaughtered. During the initial four years of the ban therapeutic usage of antimicrobials increased and overall antibiotic consumption increased as well. However, later studies have indicated these problems have been controlled with the implementation of improved environmental conditions, with no loss of meat production (Wierup 2001).

In the early 1990s, doctors in Europe began to discover vancomycin-resistant *Enterococcus* (VRE) in cultures from their patients. After a search for the origin of the resistant bacteria, in 1993 the reservoir was identified as farms where avoparcin was used as a growth promoter, leading to the 1997 ban of avoparcin for all uses in agriculture. In 1999, the European Union banned the use of AGPs from drug classes that are medically important to humans, including tylosin, spiramycin, virginiamycin, and bacitracin, and implemented other disease preventing measures. Finally, in 2006 the EU banned all AGPs for use in food animals (Cogliani et al. 2011).

Denmark enacted its own ban AGPs in 1994. The legislation restricted the direct sale of therapeutic antimicrobials from veterinarians and banned routine prophylactic use of antimicrobials. In 1995 the Danes set up the Danish Integrated Antimicrobial Resistance Monitoring and Research Program (DANMAP). DANMAP was put into place to measure the impact of withdrawing AGPs in farm animals and monitor antibiotic resistance. Since the bans antimicrobial use has decreased by 51% in swine and 90% in poultry. Production of pork since the ban is up by 47% and production of poultry has increased slightly. Therapeutic usage of antimicrobials has risen since the ban due to outbreaks of post-weaning multisystemic wasting syndrome and *Lawsonia intracellularis* in swine. However, the overall use of macrolides was reduced; a drug class deemed critically important for human medicine by the World Health Organization. The Danes have noted a significant reduction in macrolide resistance and avilomycin resistance in *E. faecium* among chickens, as well as reductions in vancomycin-resistant *E. faecium* in chickens and pigs (Cogliani et al. 2011).

The case of the Netherlands tells a different story. The Dutch introduced MARAN in 1999 to monitor antibiotic resistance and took part in the EU ban on AGPs in 2006. However, with the ban on AGPs, therapeutic use of antibiotics increased, keeping the overall usage the same. The Dutch had insufficient government control over farming practices, antibiotic sales and use, and infection control measures. Throughout farms in the Netherlands there is high prevalence of multidrug-resistant bacteria. Still, with the ban of AGPs, the Netherlands has reported a decrease in VRE among food animals and a decrease resistance to avilamycin. The Dutch example demonstrates that with bans on AGPs requires other interventions as well to prevent increased inappropriate therapeutic usage of antimicrobial agents (Cogliani et al. 2011).

The FDA's Struggle to Reduce Antibiotic Use in Animal Feed in the United States

Antimicrobials received approval for use as growth promoters in food animal production in the early 1950s. By the late 1960s, after an epidemic of infection caused by resistant strains of *Salmonella typhimurium*, concerns about the safety of this practice quickly spread from the United Kingdom to Sweden, Denmark, and the United States, among other countries. The Swann Report, presented in 1969 by a Joint Committee in the UK, reported the findings of an investigation into the use of antibiotics in animal food production and veterinary medicine. Not only did this report conclude that bacterial resistance "resulted from the use of antibiotics for growth promotion", but also that the resistance may be transferred to other bacteria and ultimately to the enteric bacteria of humans (FDA 2012).

In response to the concerns raised in this report, the FDA established its own taskforce in 1970 to review the use of antibiotics in animal feed. The task force consisted of specialists in infectious disease and animal science from the FDA, NIH, US Department of Agriculture, CDC, as well as consultants from universities and food industry. Like their European counterparts, the report agreed that the use of antimicrobials in food-producing animals was associated with the development of resistant bacteria, which may ultimately result in untreatable human disease. Additionally, the task force included recommendations to limit the use of medically important antimicrobials for growth promotion.

As a result of these reports, the European Union approved a ban of tetracycline, penicillin, and streptomycin for growth promotion in the early 1970s. Led by a push by its farmers, Sweden followed in 1986 with a ban on all antibiotics for grown promotion (Cogliani et al. 2011). The recommendations made by the FDA, on the other hand, failed to produce significant change in the animal-production industry in the United States.

In 1977, the FDA attempted to ban the use of penicillin and tetracycline in animal feed. This proposal was met with opposition from Congress, who directed the FDA to conduct further studies to demonstrate that drug-resistant bacteria originating in animals caused serious disease in humans. The study, conducted by the National Academy of Sciences, concluded that the current research was limited and poorly constructed, and that further studies were needed to prove the connection between antimicrobial use in animal feed and human infection by resistant organisms.

In an attempt to demonstrate this association, the CDC and FDA worked with the Seattle-King County Health Department in 1984 to evaluate the relationship between *Salmonella* species and *Campylobacter jejuni* in animal foods, and human illness caused by those organisms. Over a 20-month period, the study was able to identify *C. jejuni* in humans that originated in poultry, and demonstrated similar antibiotic resistance patterns in isolates from poultry and humans (Seattle-King County Department of Public Health 1984). Numerous studies have since supported the conclusion that bacterial pathogens and antibiotic resistance do in fact move between animals and humans via the food production system.

Since this landmark study, subsequent reports have been published by international organizations, including the FDA, the World Health Organization, the Institute of Medicine, the National Research Council, the United States Government Accountability Office, the World Organization for Animal Health, and the Food and Agriculture Organization of the United Nations. These reports emphasized that "low-level, long-term exposure to antimicrobials may have greater selective potential than short-term, full-dose therapeutic use (WHO 1997)," and implicated their use for growth promotion as a potential human health hazard (IOM 1989). A 2003 report by the Institute of Medicine and the National Academy of Sciences recommended that the FDA ban the use of medically important antimicrobials for the purpose of growth promotion in animals.

Despite the growing body of evidence that accumulated over 40 years since initial concerns were raised over the use of antimicrobials for growth promotion, the FDA struggled to enact significant change. Pharmaceutical companies that produced antimicrobial drugs, as well as the farmers who utilized them, had little incentive to follow the recommendations issued by the FDA. The elimination of antimicrobials from use in animal food production represents a perceived threat to the economic viability of farms, which fear a decline in efficiency and productivity. While this concern remains in the United States, long-term studies in Denmark have shown that pork and poultry production actually increased following the 2000 ban on the use of antimicrobials for growth promotion and disease prevention (Aarestrup et al. 2010).

The Future of Antibiotic Use in Food-Producing Animals in the United States

Despite mounting evidence and increased awareness that antibiotic use for growth promotion poses a threat to human health, there has been minimal legislation to curb the practice. The discrepancy between scientific consensus and public policy stems from industry influence and biases that suggest the threat of antibiotic resistance is overstated.

In 2008, the Pew Commission on Industrial Farm Animal Production released a report that outlined the barriers and recommendations to address the growing problems associated with industrial animal food production. The Commission found "significant influence by the industry at every turn: in academic research, agricultural policy development, and government regulation and enforcement (Pew 2008)."

Supporters of the use of antibiotics in animal feed as growth-promoters continue to argue that there is not enough scientific evidence to support major changes in the industry. Some contend that the use of antibiotics for growth promotion has minimal effect on the development of antibiotic resistance (Wallinga and Burch 2013), while others believe that the practice has little or no effect on human health (Turnidge 2004). The predominant argument against regulations to limit the use of antibiotics for growth promotion remains the possibility of a decline in production. Farmers and lobbyists contend that the rate of growth would decrease, while the incidence of serious infection and death among livestock would increase. This belief persists despite experiences in European countries, where the WHO reported little or no adverse impact on production or efficiency as a result of the ban on antibiotic growth promoters (Kjeldsen 2007, PEW 2010).

Over the past decade, there has been cause for optimism that the scientific community may finally be gaining an edge in the battle over public policy. The failure to effect significant change has resulted in an increasingly vocal movement that supports increased leadership, organization, evidence-based decision making and health systems reform to address the problem of antibiotics use in animal food production (Jay 2010).

Representative of this momentum, in March 2013, the Preservation of Antibiotics for Medical Treatment Act (PAMTA) was proposed to congress. The bill, which sought to increase oversight of antibiotic use on industrial farms, was supported by federal agencies, including the FDA and National Academy of Sciences, as well as medical organizations, including the American Medical Association, American Academy of

Pediatrics, and Infectious Diseases Society of America. PAMTA required that drug manufacturers prove antibiotics used for growth promotion did not adversely affect human health. This represented a major shift in the evaluation of antibiotic use, where previously drugs could be used *unless* they were proven to harmful (PEW 2010). While PAMTA ultimately failed to receive congressional approval, the introduction of the legislation represented a major step towards limiting antibiotic use in industrial animal farming.

In December 2013, the FDA released *Guidance for Industry #213*, which included strong recommendations to finally address the use of antibiotics as growth promoters. The FDA reiterated their stance that "production indications such as increased weight gain or improved feed efficiency" are not appropriate uses for antimicrobial drugs, and appealed to pharmaceutical companies to voluntarily change the antibiotic labels by removing the indication for growth promotion (MASSPIRG 2014, FDA 2013). The report also recommended increased veterinary supervision of antibiotic use for livestock, requiring a veterinarian's approval before antibiotics are added to animal feed. Antibiotic manufacturers are expected to implement the changes voluntarily over a three-year phasing out period, after which the FDA plans to re-evaluate the rate of adoption of the proposed changes. In an effort to utilize the pressure of public opinion, the FDA website lists all antimicrobial products affected by GFI #213, as well as summary information regarding manufacturer cooperation (FDA 2013).

While the failure of PAMTA was a mild disappointment, early outcomes of the FDA recommendations in GFI #213 are cause for optimism. All twenty-six antibiotic manufacturers affected by the guidelines have agreed to comply (FDA 2014), despite the lack of legal requirement to do so. It remains to be seen, however, if these changes will result in significant impact on the quantity of antibiotics used for animal food production.

Critics of the FDA's recommendations question the strength of the guidelines, citing examples from European legislation as proof that partial bans fail to curb antibiotic consumption. Following the ban of antibiotics for growth promotion in the European Union, several countries reported an increased use of antibiotics for 'disease prevention', with little to no change in total antibiotic utilization (Netherlands Ministry of Economics Affairs 2014). In response, the Netherlands enacted regulations—supported by the country's farmers—that called for a 70 percent decrease in antibiotic use by 2015. In the five years since these regulations were introduced, antibiotic consumption in industrial animal farming has dropped by more than 50 percent (Mevius and Heederik 2014).

There is legitimate concern that the new FDA guidelines will result in a similar shift of antibiotic use, without affecting overall usage. Factory farms can simply claim the drugs are for disease prevention (MASSPIRG 2014). Nonetheless, recent events suggest that the tide is turning in the effort to reduce antibiotic use for non-therapeutic purposes in animal food production.

Conclusion

There is a clear consensus among international science and health organizations that the long-term, low-dose use of antibiotics in animal food production is a major contributor to the growing problem of multi-drug resistant organisms. In the 70 years since Alexander Fleming warned of the risk of antibiotic resistance, an overwhelming body of evidence has supported his assertion. Multi-drug resistant organisms are responsible for an alarming number of deaths in the United States, and have made common infections increasingly difficult to treat. And while efforts are underway to address the problem of inappropriate prescribing in human medicine, the irresponsible use of antibiotics as growth promoters has largely gone unaddressed since the practice began in the 1950s.

Recent events, however, have given scientists and healthcare providers reason for optimism. From 2000 to 2008, annual NIH funding for research related to food animal production and human health increased from $0.2 million to $5.9 million (JHCLF 2013). Although research funding has since plateaued, the increase signals a heightened awareness and concern for this threat to public health (Pew 2008).

Over 40 years of research has demonstrated that antibiotics used for growth promotion contribute to antimicrobial resistance in human pathogens. The solution to this problem, on the other hand, is less clear. To this end, researchers and public policy advocates are focusing on several key areas: (1) expansion of preventative medicines (e.g., vaccines) to provide alternatives to antibiotic use in animal farming; (2) legislation and health systems policies to promote infection control and prevent antimicrobial resistance; and (3) development of novel antimicrobial drugs (Jay 2010).

Clearly, the unsanitary environment prevalent among industrial animal farms also contributes to the problem through increased rates of infection, necessitating the frequent use of antibiotics. Therefore, efforts to curb the misuse of antibiotics in these facilities must include strategies to improve hygiene (Gulland 2013).

Antimicrobial agents must no longer be used for the purpose of improved feed efficiency and growth promotion in food producing animals. This practice poses an enormous threat to the health of the world's population—common, previously treatable infections are increasingly the cause of serious illness and death. Scientists and healthcare providers have struggled to raise awareness of the problem for the past five decades. With the help of positive outcomes in response to legislation in Europe, the issue is finally arriving to the national discussion in the United States. While scientific advances may provide a suitable alternative to antibiotics in the prevention of disease in farm animals, significant change can only occur with the cooperation and motivation of politicians and industry leaders.

References

Aarestrup, F.M., V.F. Jensen, H.D. Emborg, E. Jacobsen and H.C. Wegener. 2010. Changes in the use of antimicrobials and the effects on productivity of swine farms in Denmark. Am. J. Vet. Res. 71: 726–733.

Alali, W.Q., H.M. Scott, R.B. Harvey, B. Norby, D.B. Lawhorn and S.D. Pillai. 2008. Longitudinal study of antimicrobial resistance among *Escherichia coli* isolates from integrated multisite cohorts of humans and swine. Appl. Environ. Microbiol. 74(12): 3672–81.

Castanon, J.I. 2007. History of the use of antibiotic as growth promoters in European poultry feeds. Poult. Sci. 86(11): 2466–2471.

Chattopadhyay, M.K. 2014. Use of antibiotics as feed additives: A burning question. Frontiers Microbiol. Vol. 5, Article 334.

Cogliani, C., H. Goossens and C. Greko. 2011. Restricting antimicrobial use in food animals: lessons from Europe. Microbe. 6(6): 274–279.

[FDA] United States Food and Drug Administration. 2012. Guidance for industry #209: the judicious use of medically important antimicrobial drugs in food-producing animals. Published April 11, 2012. Available at: http://www.fda.gov/downloads/AnimalVeterinary/GuidanceComplianceEnforcement/GuidanceforIndustry/UCM216936.pdf.

[FDA] United States Food and Drug Administration. 2013. Guidance for Industry #213: The Judicious Use of Medically Important Antimicrobial Drugs in Food-Producing Animals. Published April 11, 2012. Available at: http://www.fda.gov/downloads/AnimalVeterinary/GuidanceComplianceEnforcement/GuidanceforIndustry/UCM216936.pdf.

[FDA] United States Food and Drug Administration. 2014. FDA secures full industry engagement on antimicrobial resistance strategy. Press release, June 30, 2014.

[FDA] United States Food and Drug Administration. 2014. Summary Report on Antimicrobials Sold or Distributed for Use in Food-Producing Animals. Published October 2, 2014. Available at: http://www.fda.gov/downloads/ForIndustry/UserFees/AnimalDrugUserFeeActADUFA/UCM338170.pdf.

Fleming, A. Penicillin, Nobel Lecture. December 11, 1945. Available at: http://www.nobelprize.org/nobel_prizes/medicine/laureates/1945/fleming-lecture.pdf.

Food and Agriculture Organization of the United Nations, World Health Organization, World Organization for Animal Health. 2003. Expert Workshop on Non-Human Antimicrobial

Usage and Antimicrobial Resistance: Scientific Assessment. Geneva. Available at: https://archive.org/stream/328496-joint-fao-oie-who-expert-workshop/328496-joint-fao-oie-who-expert-workshop_djvu.txt.

Gulland, A. 2013. Antimicrobial resistance will surge unless use of antibiotics in animal feed is reduced. BMJ. 347: f6050.

Gustafson, R.H. and R.E. Bowen. 1997. Antibiotic use in animal agriculture. J. Appl. Microbiol. 83: 531–541.

Helms, M., P. Vastrup, P. Gerner-Smith and K. Mølbak. 2002. Excess mortality associated with antimicrobial drug-resistant *Salmonella* typhimurium. Emerg. Infect. Diseases. 8: 498-495.

Holmberg, S.D., J.G. Wells and M.L. Cohen. 1984. Animal-to-man transmission of antimicrobial-resistant Salmonella: investigations of U.S. outbreaks, 1971–1983. Science. 225(4664): 833–835.

Holmberg, S.D., S.L. Solomon and P.A. Blake. 1987. Health and economic impacts of antimicrobial resistance. Rev. Infect. Dis. 9(6): 1065–1078.

House of Representatives. 2013. 1150, The Preservation of Antibiotics for Medical Treatment Act. 2013. Available at: https://www.congress.gov/bill/113th-congress/house-bill/1150.

Hughes, P. and J. Heritage. 2002. Antibiotic growth promoters. Feed Tech. 6.8: 20–22.

Institute of Medicine, Division of Health Promotion and Disease Prevention. 1989. Human health risks with the subtherapeutic use of penicillin or tetracyclines in animal feed. Available at: http://www.fda.gov/ohrms/dockets/dailys/03/Aug03/081403/03n-0324-bkg0001-05-tab-4-01-vol2.pdf.

Jay, S.J. 2010. Statement submitted for the hearing record, Antibiotic resistance and the use of antibiotics in animal agriculture. House Committee on Energy and Commerce, Subcommittee on Health. Published July 14, 2010.

Johns Hopkins Center for a Livable Future. 2013. Industrial food animal production in America: examining the impact of Pew Commission's priority recommendations. Johns Hopkins Bloomberg School of Public Health. Available at http://www.jhsph.edu/research/centers-and-institutes/johns-hopkins-center-for-a-livable-future/_pdf/research/clf_reports/CLF-PEW-for%20Web.pdf.

Kjeldsen, N.J. 2007. Consequences of the removal of antibiotic growth promoters in the Danish pig industry. Danish integrated Antimicrobial Resistance Monitoring and Research Program (DANMAP) 2007 report: 81–83.

Lee, L.A., N.D. Puhr, E.K. Maloney, N.H. Bean and R.V. Tauxe. 1994. Increase in antimicrobial-resistant Salmonella infections in the United States, 1989–1990. J. Infect. Dis. 170(1): 128–34.

Levy, S.B., G.B. Fitzgerald and A.B. Macone. 1976. Spread of antibiotic resistance plasmids from chicken to chicken and from chicken to man. Nature 260: 40–42.

Looft, T., T.A. Johnson, H.K. Allen, D.O. Bayles, D.P. Alt, R.D. Stedtfeld, W.J. Sul, T.M. Stedtfeld, B. Chai, J.R. Cole, S.A. Hashsham, J.M. Tiedje and T.B. Stanton. 2011. In-feed antibiotic effects on the swine intestinal microbiome. Proceedings of the National Academy of Sciences. 109: 1691–1696.

Mainous, A.G., W.J. Hueston, M.P. Davis and W.S. Pearson. 2003. Trends in antimicrobial prescribing for bronchitis and upper respiratory infections among adults and children. Am. J. Pub. Health. 93(11): 1910–14.

Martin, L.J., M. Fyfe, K. Doré, J.A. Buxton, F. Pollari, B. Henry, D. Middleton, R. Ahmed, F. Jamieson, B. Ciebin, S.A. McEwen and J.B. Wilson. 2004. Increased burden of illness associated with antimicrobial-resistant *Salmonella* enterica serotype typhimurium infections. J. Infect Dis. 189: 377–384.

MASSPIRG Education Fund. 2014. Weak medicine: why the FDA's guidelines are inadequate to curb antibiotic resistance and protect public health. Masspirg Education Fund. Published

August 22, 2014. Available at: http://www.masspirg.org/sites/pirg/files/reports/ Weak%20Medicine%20vMA%20%28web%29%20copy_0.pdf.

Mellon, M., C. Benbrook and K. Bebrook. 2001. Hogging it!: Estimates of Antimicrobial Abuse in Livestock. Cambridge, MA: Union of Concerned Scientists. pps. 51–53; 60.

Mevius, D. and D. Heederik. 2014. Reduction of antibiotic use in animals "let's go Dutch". J. Verbrauch. Lebensm. 9(2): 177–181.

Moore, P.R., A. Evenson, T.D. Luckey, E. McCoy, C.A. Elvehjem and E.B. Hart. 1946. Use of sulfasuxidine, streptothricin and streptomycin in nutritional studies with the chick. J. Biol. Chem. 165: 437–41.

Nelson, J.M., K.E. Smith, D.J. Vugia, T. Rabatsky-Ehr, S.D. Segler, H.D. Kassenborg, S.M. Zansky, K. Joyce, N. Marano, R.M. Hoekstra and F.J. Angulo. 2004. Prolonged diarrhea due to ciprofloxacin-resistant *Campylobacter* infection. J. Infect. Dis. 190: 1150–1157.

Netherlands Ministry of Economic Affairs. 2014. Reduced and responsible: policy on the use of antibiotics in food-producing animals in the Netherlands. The Hague, The Netherlands: Ministry of Economic Affairs, Government of the Netherlands. Published February 1, 2014. Available at: http://www.government.nl/issues/antibiotic-resistance/documents-and-publications/leaflets/2014/02/28/reduced-and-responsible-use-of-antibiotics-in-food-producing-animals-in-the-netherlands.html.

Oliver, S.P., S.E. Murinda and B.M. Jayarao. 2011. Impact of antibiotic use in adult dairy cows on antimicrobial resistance of veterinary and human pathogens: a comprehensive review. Foodborne Pathog. Dis. 8: 337–355.

Pew Commission on Industrial Farm Animal Production. 2008. Putting Meat on the Table: Industrial Farm Animal Production in America. Baltimore, Maryland: Johns Hopkins Bloomberg School of Public Health. Available at: http://www. ncifap.org/_images/ PCIFAPFin.pdf.

PEW. 2010. Antibiotic resistant bacteria in animals and unnecessary human health risk. Published February 8, 2010. Available at: http://www.pewtrusts.org/en/research-and-analysis/issue-briefs/2010/02/08/antibioticresistant-bacteria-in-animals-and-unnecessary-human-health-risks.

PEW. 2010. Avoiding antibiotic resistance: Denmark's ban on growth promoting antibiotics in food animals. Published November 1, 2010.

PEW. 2014. Overuse of Antibiotics in Food Animal Production: Science Fact Sheet. Accessed online Jan 18, 2015 at: http://www.pewtrusts.org/~/media/Assets/2014/06/Overuse_ Science_Backgrounder_v3.pdf?la=n.

Seattle-King County Department of Public Health. 1984. Surveillance of the flow of salmonella and campylobacter in a community. 1–388.

Smith, H.W. 1968. Antimicrobial drugs in animal feeds. Nature. 218: 728–31.

Spellberg, B., J.G. Bartlett and D.N. Gilbert. 2013. The future of antibiotics and resistance. N. Engl. J. Med. 368: 299–302.

Stanton, T.B., S.B. Humphrey and W.C. Stoffregen. 2011. Chlortetracycline-resistant intestinal bacteria in organically raised and feral swine. Appl. Environ. Microbiol. 77(7): 167–7170.

The Humane Society of the United States. Farm Animal Statistics: Slaughter Totals, U.S. Slaughter Totals 1950–2014. 2014. Accessed Jan 18, 2015 at: http://www.humanesociety. org/news/resources/research/stats_slaughter_totals.html.

Turnidge, J. 2004. Antibiotic use in animals—prejudices, perceptions and realities. J. Antimicrob. Chemother. 53: 26–27.

Wallinga, D. and D.G. Burch. 2013. Does adding routine antibiotics to animal feed pose a serious risk to human health? BMJ. 347: f4214.

Wierup, M. 2001. The Swedish experience of the 1986 year ban of antimicrobial growth promoters, with special reference to animal health, disease prevention, productivity, and usage of antimicrobials. Microb. Drug Resist. 7(2): 183–190.

[WHO] World Health Organization. 1997. The medical impact of antimicrobial use in food animals. Available at: http://whqlibdoc.who.int/hq/1997/WHO_EMC_ZOO_97.4.pdf.

CHAPTER 9

Role of Probiotics and Prebiotics in the Management of Obesity

K.K. Kondepudi,[1,*] *D.P. Singh,*[1] *K. Podili,*[2]
R.K. Boparai[3] *and M. Bishnoi*[1,*]

Introduction

Obesity is one of the most visible public health problems that affect virtually all age and socio-economic groups. In 2014, WHO estimated around 2 billion overweight people around the globe, out of which 600 million (13% of the world's adult population (11% of men and 15% of women) people are obese (WHO 2014b). In developed countries obesity has already reached pandemic proportions (In USA, 66–70% of population is overweight or obese) and in other countries it is increasing at an alarming rate based on the World Health Statistics 2012 Report, WHO (WHO 2012). In developing countries with emerging economies (classified by the World Bank as lower- and middle-income countries, i.e., China, Brazil, India, and Mexico) the rate of increase in obesity has been more than that of developed countries. Many of these countries are now facing a double nutritional burden of disease. Besides problems related to under-nutrition and resultant

---
[1] National Agri-Food Biotechnology Institute (NABI), SAS Nagar, Punjab, India.
[2] Division of Biomedical Sciences, School of Biosciences and Technology, Vellore Institute of Technology (VIT), Vellore, Tamil Nadu, India.
[3] Department of Biotechnology, Government College for Girls-Sector 42, Chandigarh.
* Corresponding authors: kiran@nabi.res.in; mbishnoi@nabi.res.in

increase in infectious diseases, they have to deal with the rapid increase in non-communicable diseases like obesity, diabetes and hypertension (WHO 2014b). Juvenile obesity is on the rise too (Ogden et al. 2014). Children in underdeveloped and developing nations are more vulnerable to inadequate pre-natal nutrition and high intake of dense energy and low nutrient food. These dietary patterns and limited physical activity have resulted in obese but under-nourished children. In 2013, 42 million children under the age of 5 were overweight or obese. Childhood obesity is associated with a higher chance of developing life style disorders, premature death and disability in adulthood (WHO 2014a). In addition to increased risks in future, obese children experience breathing difficulties, increased risk of fractures, hypertension, and early markers of cardiovascular disease, insulin resistance and psychological effects. The overall discouraging scenario of increase in adult and childhood obesity is linked to simultaneous increase in other non-communicable complications like cardiovascular diseases (mainly heart disease and stroke), which were the leading cause of death in 2012; diabetes; musculoskeletal disorders (especially osteoarthritis—a highly disabling degenerative disease of the joints); some cancers (endometrial, breast, and colon). Thus, the situation requires urgent plans and strategies for the management of obesity and associated complications (Ng et al. 2014).

Does Obesity Need Drug Based Treatment?

Continuous scientific efforts have been made till to date, but effective and safe pharmacological interventions for the prevention or treatment of obesity are not available. Current anti-obesity medications are pharmacological agents which can reduce or control weight by affecting one of the fundamental processes of the weight regulation in human body, i.e., altering satiety or hunger, metabolism or consumption of calories. As, these drugs alter one of physiological mechanisms of the body, they are always prone to cause many side effects (Nathan et al. 2011, Baboota et al. 2013). Alarmingly, many of these novel anti-obesity drugs have not achieved the level of clinical effectiveness required by regulatory authorities, while few that are effective that are associated with severe adverse side effects that limit their long term use. In the 1990s anti-obesity drug fen-phen (combination of fenfluramine and phentermine) was approved by FDA. It was associated with pulmonary hypertension and heart valve problems which led to its withdrawal. Rimonabant (endocannabinoid receptor blocker) and sibutramine (catecholamine metabolism inhibitor) were also approved in the following years but withdrawn due to cardiovascular and psychological side effects. Pancreatic lipase inhibitor orlistat (approved in 1999) is also associated with multiple side effects like steatorrhoea, fecal incontinence, flatulence and malabsorption of fat-soluble vitamins along with increased incidence of liver and kidney diseases. Another drug lorcaserin (Belviq), a 5-hydroxytryptamine agonist, was approved by FDA in 2012 pending long term clinical trial for cardiovascular effects. Even in short term clinical trial, drug related carcinogenic effects, headache, dizziness and valvulopathy were reported. Combination of phentermine and topiramate (Qsymia), approved by FDA in 2012, also has potential teratogenic and cardiovascular side effects (Baboota et al. 2013). The number and diversity of these side effects indicate that the pathways targeted so far are important in several physiological mechanisms. A combination of nor-adrenaline and dopamine reuptake inhibitor (bupropion) and the selective opioid antagonist (naltrexone) have been approved by European drug authorities (Joo and Lee 2014). Liraglutide, the injectable, once-daily glucagon-like peptide-1 (GLP-1) analogue has been recommended by FDA (2014) for use in the treatment of obesity (Trujillo et al. 2015). Although, an overall efficacy and safety of recently approved drugs has favorable benefit/risk ratio, looking deeper into their mechanisms of action we suspect many side effects with long term use. This warrants careful analysis and sub-analysis to reveal unwanted effects of these drugs, but at the same time we have to be careful that we should not generalize all anti-obesity medications and think that each has side effects which can lead to withdrawal. Overall, the dismal history of anti-obesity medications suggests non-reliability on the medications and must look for alternative ways. Over the years it has been seen that the best and most effective options for overweight and obese individuals remain diet modification and physical exercise. It is more physiological to provide dietary regulations for prevention of life style related medical problems rather than to search for pharmacological interventions.

Pathogenesis of Obesity

Adipose Tissue: Function and Secretome

Here, we will briefly discuss the process of adipogenesis, genes and factors altering it and its association with inflammation. Adipogenesis involves division, clonal expansion and differentiation of preadipocytes into mature adipocytes (Poulos et al. 2010). During adipocyte differentiation, changes occur in cell morphology such as rounding up of the fibroblast-like preadipocytes and increased expression of several transcriptional factors and genes that are specific to the adipocyte phenotype (Niemelä et al. 2008). During adipogenesis, lipid-laden droplets begin to accumulate in the cytoplasm, which over a period of time become large and often coalesce into one or a few major droplets. Morphologically and functionally, two distinct types of adipose tissues namely White (WAT) and brown (BAT) adipose tissue with different biological roles exist in mammals (Herzig and Wolfrum 2013). WAT, represents 10% of the total body weight of a normal healthy adult, is the main site for energy storage and acts as an endocrine organ releasing adipokines also known as "adipose tissue secretome". Mature adipocytes are pivotal within adipose tissue and are well equipped to perform both hormonal (e.g., insulin) and sympathetic (e.g., adrenergic) functions. Positive energy balance leads to lipid deposition as neutral triglycerides in the WAT with the help of lipogenic enzymes. Negative energy balance, causes its hydrolysis by lipases into free fatty acids and glycerol. These are circulated via blood to peripheral tissues such as brain, liver, muscle and brown adipose tissue (Bonet et al. 2013). BAT is specific for energy dissipation and thermoregulation. In BAT, fatty acid oxidation releases energy as heat through a mitochondrial "uncoupling" phenomena (Barneda et al. 2013). Previously, it was thought that BAT is present in rodents and human infants. It was believed to be a mammalian adaptation to hypothermia as smaller animals and infants usually have larger surface area to volume ratio and are at risk of hypothermia. Recent literature about the presence and distribution of BAT both constitutive and inductive in human adults has opened up new avenues and given altogether a new perspective and direction to obesity related research (Bonet et al. 2013, Whittle et al. 2013, Fenzl and Kiefer 2014).

Adipsin (complement factor D) and Tumour Necrosis Factor-α (TNF-α) are the major proinflammatory mediators released from adipose tissue, providing evidence for a functional link between obesity and inflammation (Rosen et al. 1989). However, discovery of leptin, a satiety factor produced predominantly by adipose tissue, highlights the importance of adipose tissue as an "endocrine organ" (Murtaza et al. 2014). Leptin sends signals from adipocytes to the hypothalamus and controls appetite and energy balance. Adiponectin is another crucial hormone produced exclusively by the adipocytes. Adiponectin has both inflammatory and anti-inflammatory activities. The expression and circulating levels of adiponectin are decreased under obese conditions. Besides leptin and adiponectin, other secretory signal molecules of the adipocyte secretome, known as 'adipokines' (or 'adipocytokines'), includes pro- and anti-inflammatory cytokines, growth factors and proteins of the alternative complement system; proteins regulating blood pressure, vascular homeostasis, lipid metabolism, glucose homeostasis and angiogenesis. Balanced production of adipokines is a product of normal adipocyte function, which in turn is under nutritional influence (Stojsavljevic et al. 2014). More detailed information on adipocyte life cycle, adipose tissue associated secretome and its physiological functions, are given in Sethi and Vidal-Puig (2007) and Ouchi et al. (2011).

Obesity and Gut Microflora (GMF)

GMF is now recognized as a "second brain or microbial organ" consisting of 10^{13}–10^{14} bacterial cells, which is 10 times the amount of human cells. Its microbiome is 100 times more than the host's genome (Consortium 2012, Segata et al. 2013). GMF exists in mutualism and influences the host's physiological processes such as gut motility, gut barrier homeostasis, absorption of nutrients, energy homeostasis and fat distribution (Bondia-Pons et al. 2015, Ohno 2015). Alteration in GMF is one of the contributory factors to obesity. GMF maximizes the energy harvesting phenomena (Bäckhed 2014, Moran and Shanahan 2014). Fecal transplantation of the germ free mice with fecal microflora from obese mice made the former more obese than those transplanted from the lean counter parts (Ellekilde et al. 2014). Healthy human and animals guts have beneficial/commensal microflora that outplays the pathogenic ones. GMF metabolic function helps to maintain the metabolic health of humans and animals (Delzenne et al. 2011). Commensal microbes have a remarkable metabolic ability to metabolize endo and xenobiotics, ferment nutrients and secrete bioactive molecules that affect the host's physiology and metabolism (Delzenne et al. 2011).

Human gut metagenomic studies revealed *Firmicutes* and *Bacteroidetes* as the dominant phyla whereas *Proteobacteria*, *Fusobacteria*, *Verrucomicrobia* and *Actinobacteria* are the sub-dominant ones (Tilg and Kaser 2011). Lactobacilli, bifidobacteria and butyrate producing bacteria such as *Roseburia* and *Faecalibacterium prausnitzii* are some of the health promoting bacteria in the gut (Ventura et al. 2014, Paoli et al. 2015). Both commensal and pathogenic bacteria and their metabolites/derivatives are recognized by pattern recognition receptors (PRRs), which in turn are mediated via Toll like receptors (TLRs) and by nucleotide-binding oligomerization domain/caspase recruitment domain isoforms (NOD) receptors. Signals from commensal bacteria are essential for intestinal barrier function and gut tissue repair. Several other microbes associated molecular patterns (MAMPs) influence the normal development and function of the mucosal immune system (Rakoff-Nahoum et al. 2004, Mazmanian et al. 2005, O'Hara and Shanahan 2006) by fine-tuning the T-cell repertoires and T-helper (Th)-cell type 1 or type 2 cytokine profiles (Cebra 1999, O'Hara and Shanahan 2006). Perturbations in GMF composition arise due to antibiotics, chronic high fat diet consumption, sedentary life style, and excess intake of energy dense foods containing refined sugars. This results in the predominance of pathogenic bacteria as observed in obese, overweight and/or type 2 diabetes; nonalcoholic fatty liver disease (NAFLD) and cancer patients. Pathogenic bacteria secrete proinflammatory metabolites, such as lipopolysaccharides (LPS), which are signaled by TLR4; peptidoglycans and lipotechoic acids by TLR2 and flagellin by TLR5. These proinflammatory metabolites gain access into bloodstream due to dampened gut-barrier function thereby promoting global low-grade inflammation and metabolic endotoxemia. Obesity associated GMF dysbiosis revealed a low occurrence of *Bacteroidetes*, more *Firmicutes* and higher proportions of LPS secreting Gram negative colonizing pathogens in the gut (Ouchi et al. 2011). LPS from *Escherichia coli*, an endotoxin from *Enterobacter cloaca* B29 and peptidoglycans from other bacteria have recently been reported to cause obesity (Cani et al. 2007, Fei and Zhao 2013). Recent studies have shown that commensal microflora can counteract weight gain and metabolic endotoxemia (Anhe et al. 2014, Joyce et al. 2014). This has been attributed to the regulation of host's energy balance through cholesterol, bile acid, carbohydrate metabolisms; gut hormone secretion; immunomodulation; neurotransmitter secretion; vitamin secretion and also through epigenetic regulations in the host tissues (Figure 1).

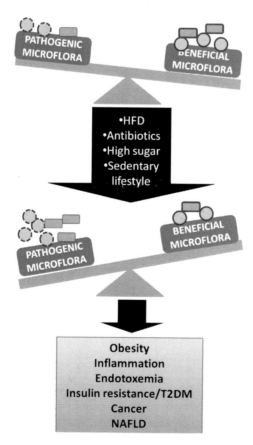

Figure 1. Gut microflora and its role in regulating host's metabolic processes.

GMF Modulation (Non-digestible carbohydrates, probiotics, synbiotics, bioactive/dietary constituents) for the Weight and Obesity Management

Non-digestible Carbohydrates (NDC) and Prebiotics

Dietary fiber (DF) is "the *edible parts of plants or analogous carbohydrates that are resistant to digestion and absorption in the human small intestine with complete or partial fermentation in the large intestine*" to simple sugars and short chain fatty acids (SCFAs). DF exerts laxation, lowers blood cholesterol and lower blood glucose levels. Further, they decrease the energy release from foods as they bulk the foods. Insoluble fibers cause excretion of bile acids with a decreased intestinal transit time. Soluble viscous fibers slow down the glucose absorption and are good substrates for microbial fermentation in the large bowl (Charalampopoulos et al. 2002, Wong et al. 2006, Cummings and Stephen 2007). Pectins, β-glucan, xylan, arabinoxylan, inulin, resistant starch and gum Arabic are some of the well-studied DFs and several studies have shown that DF consumption enhances weight loss significantly (Slavin 2013).

Enzymatic hydrolysis of the dietary fibres or by transglycosylation reactions of monomer sugars generates oligosaccharides also known as prebiotics. They are also non-digestible by the human digestive enzymes but are fermented by the gut microflora yielding SCFAs. Approved prebiotics include fructooligosaccharides; galactooligosaccharides and lactulose (Baboota et al. 2013) and recently isomalto-oligosaccharides by European Food Safety Authority (EFSA) (Tetens 2009). Other prebiotic oligosaccharides are being studied more recently but their role in promoting human gut health is yet to be established (Kondepudi et al. 2012). Various dietary fibres and prebiotics are used as dietary interventions in preclinical and clinical models of obesity and related complications and the outcome of these studies are shown in Table 1.

Probiotics

Probiotics are the "*live microorganisms that when administered in adequate amount confer health benefits to the host*" (Sanders 2008). Probiotics are believed to promote gastrointestinal health as they are known to improve the host's immune system, nutrient bioavailability and alleviate lactose intolerance (Naïma et al. 2014). Probiotic bacteria, their secondary metabolites and bacterially derived molecules, e.g., MMAPs, are gaining importance in regulating the host metabolic processes, weight gain and obesity.

Table 1. GMF modulation (Non-digestible carbohydrates/prebiotics and synbiotics) for weight maintenance and obesity management.

Prebiotics/Synbiotics	Effects	Mechanisms	References
PREBIOTICS			
Whole grains plus traditional medicinal foods rich in dietary fibers (adlay, oat, buckwheat, white bean, yellow corn, red bean, soybean, yam, big jujube, peanuts, lotus seeds, wolfberry) OR prebiotics (fructo-oligosaccharides or oligoisomaltose) OR soluble prebiotics (guar gum, pectin, konjac flour, resistant starch, hemicelluloses)	Improved insulin sensitivity, lipid profile and blood pressure along with reduced weight gain in centrally obese people: a self-controlled clinical trial	Reduced endotoxin producing bacteria and gut barrier-protecting bacteria. Reduced inflammation and gut permeability	(Xiao et al. 2014)
Insulin type fructans and related oligosaccharides	Decreases in appetite, fat mass and hepatic insulin resistance	Improved gut barrier function via GLP-2 and satiety effect via GLP-1 release	(Cani et al. 2006, Cani et al. 2009)
Non-digestible fermentable carbohydrate	Increased satiety and reduction in adiposity	Hypothalamic control of feed intake via SCFAs production	(Frost et al. 2014)
Fructooligosaccharides (FOS) enriched cookies	Improved satiety without affecting cardiovascular risk factors in obese population	Parallel randomized controlled trial (RCT)	(de Luis et al. 2013)
Insulin/oligofructose 50/50% mix	Insulin type fructans selectively change gut microbiota in obese women and reduced metabolic endotoxemia in a double blinded placebo controlled trial		(Dewulf et al. 2013)
Trans-galactooligosaccharides	Improved beneficial gut microbiota abundance, immune response, insulin, triglycerides (TG), total cholesterol (TC) concentration in obese people with metabolic syndrome in a cross-over RCT		(Vulevic et al. 2013)

	Antiobesity effect, cardio-protective effect		For review (El Khoury et al. 2012)
β-glucans			
Oligofructose	Improved metabolic outcomes in diet induced obesity in male Sprague Dawley rats		(Pyra et al. 2012)
Swedish brown beans (*Phaseolus vulgaris* var. *nanus*)	Improved glycemic control, increased PYY and GLP-2, reduced ghrelin	Potential outcomes via modulation of glucose dependent insulinotropic polypeptide, leptin and phosphorylated acetyl CoA carboxylase and gut microflora (increased Bifidobacteria and reduced Clostridium leptum)	(Nilsson et al. 2013)
Wheat dextrin	Increased short time satiety and decreased caloric intake in overweight adults in double blind placebo controlled RCT	Improved colonic fermentation of brown bean food may beneficially affect the metabolic outcomes	(Guérin-Deremaux et al. 2011)
Wheat arabinoxylan	Improved gut barrier function and lower circulating inflammatory markers and adiposity via modulation of various genes involved in adipocytes differentiation, fatty acid uptake and fatty acid oxidation	Increased Bifidobacterium animalis lactis, *Bacteroides-Prevotella* spp. and *Roseburia* spp.	(Neyrinck et al. 2011)
Wheat derived arabinoxylan oligosaccharides	Control obesity and low grade systemic inflammation in mice fed on high fat diet (HFD)	Bifidogenic effect along with an increased satitogenic peptides in colon	(Neyrinck et al. 2012)
Inulin	Improved glycemic control and systemic inflammation in type 2 diabetic females in a RCT	Modulation of inflammation and metabolic endotoxemia	(Dehghan et al. 2014)
Arabinoxylan	Improved glycemic control and lipid profile in RCTs	Effective in hyperglycemia control but no effects on adipokines	(Garcia et al. 2006)

Table 1. contd....

Table 1. contd.

Prebiotics/Synbiotics	Effects	Mechanisms	References
Resistant starch/Arabinoxylans/pectin	Improved gut microbiota, reduced adiposity, improved insulin sensitivity, improved lipid profile		For review (Kondepudi et al. 2014)
Insulin-type fructans (ITF)	Antiobesity effect in diet induced obesity in C57BL/6 mice	Decreased PPAR-γ activated process and modulate gut microbial abundance	(Dewulf et al. 2011)
Insulin type fructans	i) Consumption of (ITF) modulates Bifidobacterium sp. parallel placebo controlled RCT in obese women ii) Counteract peroxisome proliferator-activated receptors gamma type (PPARγ) activators (thiazolidinedione) associated weight gain without affecting its beneficial effect on glucose homeostasis in HFD fed rats	i) Reduction in fecal SCFAs concentration that can lessen the risk factors associated with it ii) Promote brite like phenotype in visceral adipose tissue	(Salazar et al. 2014) (Alligier et al. 2014)
Resistant starch from high amylase maize	Resistant starch alone or in combination to dietary butyrate reduces abdominal fat	This action was supposed to be via increased serum PYY and GLP-1	(Vidrine et al. 2014)
	Reduced body fat in ovariectomized (OVX) rats	OVX caused significant weight gain which was attenuated by modulation of gut microbiota	(Keenan et al. 2013)
Fucoidan	i) Inhibition of fat accumulation in 3T3-L1 cell lines ii) Inhibit pancreatic lipase *in-vitro* iii) Reduced plasma TG level in rats iv) Reduced body weight gain		For review (Vo and Kim 2013, Kondepudi et al. 2014)

Agent	Effects	Mechanism/Type	Reference
β-glucans	i) Reduction in hunger ii) Reduction in energy intake, plasma ghrelin and increased PYY		For review (Kondepudi et al. 2014)
Gum Arabic	Reduced body weight gain and body fat percentage in healthy females in double blind placebo controlled trial		(Babiker et al. 2012)
Finger millet bran	Prevented body weight gain, improved lipid profile and anti-inflammatory status, alleviated oxidative stress in male LACA mice	Via modulation of gut microflora (Lactobecillus, Bifidobacteria, Akkermansia, Bacteriodetes, Enterobacter, Firmicutes) and PPARγ, C/EBPα like transcription factor	(Murtaza et al. 2014)
Resistant dextrin	Prevent insulin resistance and systemic inflammation in type 2 diabetic women in a RCT	Improved HOMA-IR index, IL-6, TNF-α, LPS concentration	(Aliasgharzadeh et al. 2015)
SYNBIOTICS			
L. sporogenes, 1 × 10⁷ colony forming units (CFU) + 0.04 g insulin (three times in a day)	Decreased serum insulin and high sensitivity C-reactive protein (hs-CRP) tending to have significant effect on TG, low density lipoprotein (LDL-c), homeostatic model assessment for insulin resistance (HOMA-IR), and high density lipoprotein (HDL) in diabetic patients. It also improves plasma total reduced glutathione (GSH) concentration	Cross-over RCT	(Asemi et al. 2014)
B. longum with FOS	Improved glucose/lipid profile and reduced Nonalcoholic steatohepatitis	Parallel RCT	(Malaguarnera et al. 2012)

Table 1. contd....

Table 1. contd.

Prebiotics/Synbiotics	Effects	Mechanisms	References
L. casei + L. rhamnosus + S. thermophilus + B. breve + L. acidophilus + B. longum + L. bulgaricus (4 × 108) + FOS	Decreased fasting insulin and fasting glucose in obese patient with NAFLD	Parallel RCT	(Eslamparast et al. 2014)
B. animalis subsp. lactis BB-12 (BB-12) + oligofructose (OFS)	Better glycemic control than individual components in diet induced male SD rats	A symbiotic action proposed	(Bomhof et al. 2014)
B. lactis HN019 + FOS, B. lactis HN019 + GOS, B. lactis HN019 + inulin, L. rhamnosus HN001 + FOS, L. rhamnosus HN001 + GOS, and L. rhamnosus HN001 + insulin	All treatments showed lowered body weight gain and decreased cecal acetic acid concentration. Rats fed L. rhamnosus HN001 had enhanced colonic β-defensin 1 and mucin (MUC)-4 gene expression. All increase the MUC–4 gene expression	Modulation of gut microflora and mucin production linked to the beneficial effect of the symbiotic combination	(Paturi et al. 2015)
Clostridium butyricum MIYAIRI 588 [CBM] + insulin	Reduced adiposity and serum triglyceride level via increased peroxisomes abundance	Induces Pex11a, acyl-coenzyme A oxidase 1, and hydroxysteroid (17-β) dehydrogenase 4	(Weng et al. 2015)

This include significant research and commercial aspects (Neish 2009, Lebeer et al. 2010, Delzenne et al. 2011). *Lactobacillus* and *Bifidobacteria* are the well-studied probiotic bacteria genera as many of these bacteria are generally recognized as safe (GRAS) (Pouwels et al. 1996, Meile et al. 2008). In many of the clinical conditions including infections, autoimmune disorders and metabolic complications such as diabetes and obesity their predominance was found to be significantly lowered (Consortium 2012, Bäckhed 2014, Bondia-Pons et al. 2015). Conversely, existing literature suggests that restoration or replenishment of these bacteria by either exogenous administration or stimulation of indigenous bacteria using DFs, prebiotics and certain phytochemicals such as polyphenols is a safe approach to improve the gastrointestinal health (Baboota et al. 2013, Raman et al. 2013, Kondepudi et al. 2014). In addition to Lactobacilli and bifidobacteria, other gut microbes such as *Akkermansia muciniphila*, a mucolytic bacterium, has tremendous potential in anti-obesity bio-therapeutics (Everard et al. 2013). Abundance of *Akkermansia muciniphila* has been reported to be significantly reduced upon high-fat consumption whereas oral supplementation with viable *Akkermansia muciniphila* reverses the high-fat diet associated changes. However the safety issue of this species for human applications and technological constraints limits its usage. Instead dietary components that could enhance the abundance of *Akkermansia* such as insulin, certain polyphenols seem to be viable option (Van den Abbeele et al. 2011, Everard et al. 2013). We have recently reported an interesting observation that dietary capsaicin, a transient receptor potential cation channel subfamily V member 1 (TRPV1) agonist, alleviated the high-fat diet-induced obesity along with significant increase in the abundance of *Akkermansia*, Lactobacilli and Bifidobacteria. Further, we are investigating the mechanisms behind stimulation of *Akkermansia* upon capsaicin supplementation and its effect on diet induced obesity (Baboota et al. 2014). Several other studies are using microbes other than lactobacillus and bifidobacteria genera. Table 2 summarizes the potential of conventional and specific probiotic strains and some non-lactic acid bacteria with a probiotic potential in alleviating the progression of obesity and its related complications.

Synbiotics

Synbiotics ("syn"-together and "bios"-life) is a combination of pre and probiotics. Such combination is expected to synergistically promote host's health as compared to individual administration of pre- and probiotics. Complementary and synergistic synbiotics are the two combinations proposed by the researchers (Kolida and Gibson 2011). Complementary synbiotics contain mono or multiple probiotic bacterial strains, and one or more than one prebiotics are added irrespective of whether exogenous

Table 2. Various probiotics investigated for obesity and related disorders.

Probiotics	Effects	References
Bifidobacterium (*B.*) *animalis* ssp. *lactis* 420 (B420) strain	Reduced tissue inflammation, endotoxemia and other alterations caused by HFD in C57BL/6 mice	For review (Torres-Fuentes et al. 2015)
Lactobacillus (*L.*) *gasseri* SBT2055 (LG2055) strain	i) Reduced abdominal adiposity, inflammation in rats fed on HFD ii) Reduced obesity in adults with obese tendencies in a randomized controlled trail iii) Reduced adipocytes size in visceral white adipose tissue in SD rats	(Sato et al. 2008, Kadooka et al. 2010, Kadooka et al. 2011)
L. gasseri BNR17 strain	Prevents diet induced obesity in rats.	
B. breve B3 strain	Prevents diet induced obesity in mice.	
L. rhamnosus PL60 strain	Reduced body weight without decreasing energy intake in mice.	(Lee et al. 2006c)
Akkermansia muciniphila	Reduced HFD induced metabolic disorders and endotoxemia	(Everard et al. 2013)
L. paracasei ssp. *paracasei* F19 (F19) strain	Decreased fat storage in germ free mice via modulation of Angptl4, a circulating lipoprotein lipase inhibitor	(Aronsson et al. 2010)
L. paracasei CNCM I-4270 (LC) / *L. rhamnosus* I-3690 (LR / *B. animalis* ssp. *lactis* I-2494 (BA) individually (10⁸ cells/day)	Reduced weight gain and macrophage infiltration in epididymal fat pad, improved insulin-glucose homeostasis in HFD fed mice	(Wang et al. 2015)
VSL #3 (lactic acid bacteria mix: live freeze dried form)	Reduced risk of hepatic encephalopathy in liver cirrhosis in a RCT	(Dhiman et al. 2014)
L. acidophilus NCFM	Improved insulin sensitivity and systemic inflammation in randomized double-blind intervention	(Andreasen et al. 2010)
L. plantarum strain No. 14 (LP14)	Reduced mean adipocyte size, white adipose tissue weight, improved lipid profile in HFD fed mice	(Takemura et al. 2010)

Strain/Treatment	Effect	Reference
L. plantarum KY1032 cell extract	Decreased lipid accumulation in 3T3-L1 adipocytes via modulation of PPARγ, CCAAT/enhancer-binding protein alpha (C/EBPα), fatty acid synthase (FASN) like transcription factors	(Park et al. 2011)
L. gasseri BNR17 strain	Increased mRNA levels of fatty acids oxidation genes, lowered fatty acids synthesis, reduced leptin and insulin in high sucrose diet fed obese mice	(Kang et al. 2013)
Soy milk fermented with *L. paracasei* ssp. *paracasei* NTU 101 and *L. plantarum* NTU 102	In-vitro differentiation of preadipocyte inhibited and lipolysis increased. Leptin expression increased in HFD fed obese rats along with improved serum lipid profile	(Lee et al. 2013)
L. brevis OK56	Inhibited HFD induced weight gain, macrophage infiltration in adipose tissue, nuclear factor kappa beta (NF-kB) activation and LPS production	(Kim et al. 2015)
B. longum and its subspecies (SPM 1205/SPM 1207/SPM 1204)	Reduced body fat, inhibit endotoxemia, improved plasma lipid profile, and improved liver function	(An et al. 2011, Chen et al. 2012)
Fermented burdock diet with *Aspergillus awamori*	Elevated faecal immunoglobulin A (IgA) and mucin. Cecal SCFAs concentration, reduction in adipose tissue weight in HFD fed rats	(Okazaki et al. 2013)
Different other species and subspecies of *Lactobacillus, Bifidobacterium*	Reduced body weight improved lipid and glucose metabolism and reduction in metabolic endotoxemia in rodent model	For review (Cani and Van Hul 2015)
L. casei fermented milk	Reduced body weight and clinical parameters in serum. Increased Bacteroides and bifidobacteria in gut of HFD fed mice	(Nunez et al. 2014)
L. acidophilus La5, *Bifidobacterium* BB12, and *L. casei* DN001 enriched yogurt	Reduced weight gain in weight-loss diet and probiotic yogurt group. Synergistic effects showed were suggested due to T-cells subset specific gene expression in peripheral blood mononuclear cells among overweight and obese individuals	(Zarrati et al. 2014)

probiotic bacteria could metabolize the prebiotic or not. Therefore the prebiotic may or may not promote growth and activity of the ingested probiotic (Gallaher and Khil 1999). Synergistic synbiotics contain mono or multiple probiotics and prebiotic is chosen to specifically enhance the survival, growth and activity of the selected probiotic strain(s). That means the probiotic component must be able to metabolize the prebiotic and the latter must be able to stimulate the innate beneficial GMF and exert synergistic effect. Some of the studies where specific synbiotic combinations are tested for their efficacy to counteract weight gain, obesity and associated complications are summarized in Table 1.

Bioactive/dietary Constituents and Gut Modulation

There is a growing literature and the list of many dietary and food constituents which modulate the gut microbiota is ever increasing (Table 3). Also, beneficial GMF metabolizes many dietary constituents into bioactive metabolites that can also act as signalling ligands modulating intestinal and immunological functions (Aura et al. 2005, Comalada et al. 2005, Lee et al. 2006b, Bashmakov et al. 2011, Baboota et al. 2013). These metabolites relay the messages from gastrointestinal tract to other peripheral tissues such as liver, adipose tissue and muscle thus regulating the host's energy metabolism. Gut-brain axis also plays an important role in many diseases. Commensal bacteria and their metabolites also protect from the proinflammatory responses from the pathogenic bacteria (Cryan and O'Mahony 2011, Bercik et al. 2012). Dietary supplementation with several bioactives, majorly antioxidants can enhance the abundance of good bacteria and limit the perturbations caused by pathogenic ones (Table 3). Bacterial metabolism of these bioactives and their potential anti-inflammatory and other activity is mainly responsible for their anti-obesity effect (Vissiennon et al. 2012, Uehara 2013, Zhang et al. 2014).

Novel Mechanisms for Gut Microflora Modulation of Host Metabolic State

Cholesterol and Bile Acid Metabolism

Gut microflora modulates obesity through bile acid metabolism. In the liver, cholesterol gets converted into cholic and chenodeoxycholic acid (primary bile acids) which aid in absorption of fats and fat-soluble vitamins from the small intestine (Sayin et al. 2013) (Reddy and Watanabe 1979). Taurine and glycine conjugation with the primary bile acids allows

Table 3. Various phytochemicals/dietary constituents investigated in obesity and related disorders.

Dietary substances/ Phytochemicals	Gut microflora modulation	Effects	References
Polyphenols mixture/ individual i) Naringenin ii) Naringin iii) Hesperetin iv) Hesperidin v) Quercetin vi) Rutin vii) Catechin	a) Variable modulation of Bacteroides galacturonicus, *Lactobacillus* sp., *Enterococcus caccae, Bifidobacterium catenulatum, Ruminococcus gauvreauii, Escherichia coli* b) Absorption and metabolism dependent on gut microfloral abundance	i) More effective compared to naringin ii) No effect iii) More effective compared to hesperidin iv) No effect v) Minimum inhibitory concentration (MIC) for *R. gauvreauii, B. galacturonicus* and *Lactobacillus* sp. vi) No effect vii) No effect	(Laparra and Sanz 2010, Duda-Chodak 2012)
Capsaicin	Inhibit *Enterobacteriaceae* and Firmicutes. Promote abundance of *Lactobacillus* and *Akkermansia*	Anti-obesity effect of capsaicin via gut microbial modulation	(Baboota et al. 2014a,b)
Cranberry extract	Increases mucin degrading bacteria *Akkermansia*	Anti-obeisty effect of Cranberry extract	(Anhe et al. 2014)
p-Coumeric acid	*Clostridium perfringens, Clostridium difficile* and *Bacteroides* spp. Suppressed, while commensal anaerobes like *Clostridium* spp., *Bifidobacterium* spp. and probiotics such as *Lactobacillus* sp. were less severely affected	Bacteriostatic activity	(Lee et al. 2006a)
Catechin/Epicatechin metabolites	*Clostridium coccoides,*Eubacterium rectale group, *Bifidobacterium* spp. Increased	Bacteriostatic or antimicrobial activities	(Tzounis et al. 2008)

Table 3. contd....

Table 3. contd.

Dietary substances/ Phytochemicals	Gut microflora modulation	Effects	References
Resveratrol	i) Increased *Lactobacillus* and *Bifidobacterium* counts in feces ii) Insulin sensitivity restoration, peripheral nerve functions	i) Prebiotic effect, maintain colonic mucosa architecture, reduced body weight loss, diminished the induced anemia and systemic inflammation ii) Diabetic foot syndrome	(Larrosa et al. 2009) (Bashmakov et al. 2011)
Daidzein (Soyabean isoflavone)	Metabolized by intestinal microflora	Estrogenic and bone sparing activity	(Uehara 2013)
Quercetin	Active quercetin aglycon production in intestine via action of various gut microflora	Anti-inflammatory effect via NF-kB pathway	(Comalada et al. 2005, Zhang et al. 2014)
Cyanidine	Metabolism to cyaniding-3-glucoside an aglycone by gut microflora	Antimicrobial properties	(Aura et al. 2005)
Metabolites of kaempferol, quercetin and myricetin	Active metabolites produced either by gut microflora or extrinsically showed activity	Anxiolytic activity, anti-inflammatory activity	(Laparra and Sanz 2010, Vissiennon et al. 2012)
Plantago maxima leaves extract	Modulation of gut microbial population	Antiadipogenic, antidiabetic, antiinflammatory, antioxidant activity	(Tinkov et al. 2014)
Ginger	Modulation of *Lactobacillus* sp. at various doses without affecting coliform, clostridium and total bacteria of ileum content	Improved growth, digestibility serum chemistry and balanced gut microflora in guinea fowls	(Oso et al. 2013)
Galangin, quercetin, and fisetin	Metabolised by *Bifidobacterium adolescentis*	Reduced production of nitric oxide in cell culture system	(Kawabata et al. 2013)
Fermentation products of a mixture of anthocyanins	Enhanced the growth of *Bifidobacterium* spp. and *Lactobacillus–Enterococcus* spp.	Possible role of bacterial metabolism of anthocyanins	(Hidalgo et al. 2012)

them to be absorbed in the distal ileum and transported to the liver (Swann et al. 2011). Bacterial de-conjugation of these in the ileum and gut microbiota-induced metabolism converts them to secondary bile acids. Gut microbial association with obesity and type II diabetes has recently been attributed to the composition of bile-acid pools as a result of bile salt hydrolase activity and signalling receptors, farnesoid X receptor (FXR) and G-protein coupled receptor (GPCR) TGR5 (Nicholson et al. 2012, Sayin et al. 2013). FXR impairs glucose homeostasis whereas TGR5 promotes it (Ma et al. 2006, Thomas et al. 2009). FXR is activated by primary bile acids whereas TGR5 binds secondary bile acids (deoxycholic acid and lithocholic acid) (Watanabe et al. 2006). TGR5 signalling in enteroendocrine L-cells induces secretion of GLP-1, thereby improving liver and pancreatic function and enhancing glucose tolerance in obese mice (Thomas et al. 2009, Tremaroli and Bäckhed 2012). Bile acids are taken up from the gut and circulated throughout the body, so activation of TGR5 and FXR in peripheral organs may contribute to overall host metabolism. Activation of TGR5 in brown adipose tissue and muscle increases energy expenditure and protects against diet-induced obesity (Watanabe et al. 2006). Hence, cholesterol and bile acid metabolism plays a significant role in gut microflora based modulation of obesity and related complications.

Microbial Metabolism of Choline

The major source of choline is phosphotidyl choline, an important component of the cell membranes (Tang et al. 2013). Host tissues can synthesize choline on its own or it can get this exogenously from dietary source *via* red meat and eggs. Choline is metabolised by the gut microbial and host enzymatic activities into methylamines, trimethylamine in intestine and further to trimethylamine-N-oxide in the liver (Nicholson et al. 2012). Gut microbial dysbiosis, as observed in obesity, might allow for excess formation of these methylamines and that might contribute for NAFLD; impaired glucose homeostasis and cardiovascular diseases.

NDCs Metabolism and Role of Short Chain Fatty Acids

Gut microbial metabolism of NDCs produces acetate, propionate and butyrate, the energy rich molecules, and the predominant SCFAs in humans and animals. Colonic acetate production has been shown to suppress appetite through central hypothalamic mechanisms along with change in neurotransmitter levels (Frost et al. 2014). Propionate and butyrate but not acetate have been reported to modulate the release of major gut hormones such as GLP-1 and peptide YY (PYY) expression involved in hunger and satiety functions. These SCFAs act through GPCRs such as GPR41 (FFAR3) and GPR43 (FFAR2) (Lin et al. 2012). Butyrate and propionate inhibit weight gain through decreasing food intake whereas acetate prevents weight gain without effecting food intake and locomotor activity, suggesting increased metabolic rate or decreased absorptive efficiency (Lin et al. 2012). Butyrate also causes epigenetic regulations by inhibiting histone deacetylase and helps in intestinal epithelial cell differentiation and proliferation, and mediates other metabolic effects (Meijer et al. 2010, Berni Canani et al. 2012, Brahe et al. 2013). SCFAs can also modulate the inflammatory status. Generation of a good SCFA profile in the colon is crucial to health and that can be achieved by dietary supplementation with dietary fibres, prebiotics and combinations (Slavin 2013). Figure 2 illustrates the link between gut microflora and obesity and highlights the possible mechanisms of action.

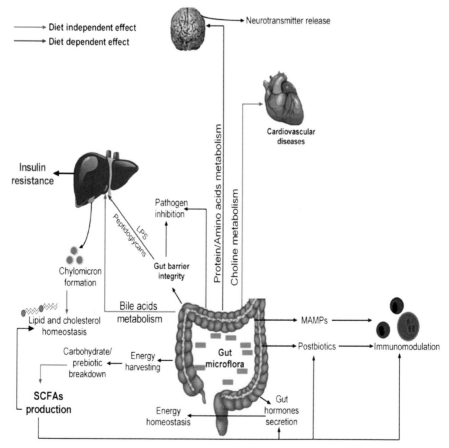

Figure 2. Gut microflora and its dysbiosis environmental factors leads to obesity, inflammation, metabolic endotoxemia, insulin resistance, NAFLD and cancers.

Conclusion

The above discussions suggest that by careful dietary manipulations one can beneficially modify the gut microflora using probiotics, prebiotics, synbiotics and plant based bioactives. However much of the available literature is based on the results obtained from interventions with probiotics, prebiotics and synbiotics using animal models of high fat or high sugar induced obesity. In order to translate the existing knowledge to human applications, there is a need for more rigorous human studies. Further, due to technical limitations of using live bacteria, research focused on bacterially derived products also known as "postbiotics" is evolving. Such molecules should be thoroughly evaluated for their efficacy and safety for the management of obesity epidemic.

Acknowledgements

Financial support from Department of Biotechnology and Department of Science and Technology, Government of India to Dr. Kondepudi K.K. and Dr. Mahendra Bishnoi is gratefully acknowledged. Authors would like to thank National Agri-Food Biotechnology Institute for providing infrastructure and fellowship to Mr. Dhirendra Pratap Singh.

Keywords: Obesity, gut microflora, dietary fibers, prebiotics, probiotics, synbiotics

References

Aliasgharzadeh, A., P. Dehghan, B.P. Gargari and M. Asghari-Jafarabadi. 2015. Resistant dextrin, as a prebiotic, improves insulin resistance and inflammation in women with type 2 diabetes: a randomised controlled clinical trial. British Journal of Nutrition. 113: 321–330.

Alligier, M., E.M. Dewulf, N. Salazar, A. Mairal, A.M. Neyrinck, P.D. Cani, D. Langin and N.M. Delzenne. 2014. Positive interaction between prebiotics and thiazolidinedione treatment on adiposity in diet-induced obese mice. Obesity (Silver Spring). 22: 1653–1661.

An, H.M., S.Y. Park, K. Lee do, J.R. Kim, M.K. Cha, S.W. Lee, H.T. Lim, K.J. Kim and N.J. Ha. 2011. Antiobesity and lipid-lowering effects of *Bifidobacterium* spp. in high fat diet-induced obese rats. Lipids Health Dis. 10:116.:10.1186/1476-1511X-1110-1116.

Andreasen, A.S., N. Larsen, T. Pedersen-Skovsgaard, R.M. Berg, K. Moller, K.D. Svendsen, M. Jakobsen and B.K. Pedersen. 2010. Effects of *Lactobacillus acidophilus* NCFM on insulin sensitivity and the systemic inflammatory response in human subjects. Br. J. Nutr. 104: 1831–1838.

Anhe, F.F., D. Roy, G. Pilon, S. Dudonne, S. Matamoros, T.V. Varin, C. Garofalo, Q. Moine, Y. Desjardins, E. Levy and A. Marette. 2014. A polyphenol-rich cranberry extract protects from diet-induced obesity, insulin resistance and intestinal inflammation in association with increased *Akkermansia* spp. population in the gut microbiota of mice. Gut. 30: 2014–307142.

Aronsson, L., Y. Huang, P. Parini, M. Korach-André, J. Håkansson, J.-Å. Gustafsson, S. Pettersson, V. Arulampalam and J. Rafter. 2010. Decreased Fat Storage by Lactobacillus Paracasei Is Associated with Increased Levels of Angiopoietin-Like 4 Protein (ANGPTL4). PLoS ONE. 5: e13087.

Asemi, Z., A. Khorrami-Rad, S.A. Alizadeh, H. Shakeri and A. Esmaillzadeh. 2014. Effects of synbiotic food consumption on metabolic status of diabetic patients: a double-blind randomized cross-over controlled clinical trial. Clin. Nutr. 33: 198–203.

Aura, A.M., P. Martin-Lopez, K.A. O'Leary, G. Williamson, K.M. Oksman-Caldentey, K. Poutanen and C. Santos-Buelga. 2005. *In vitro* metabolism of anthocyanins by human gut microflora. Eur. J. Nutr. 44: 133–142.

Babiker, R., T.H. Merghani, K. Elmusharaf, R.M. Badi, F. Lang and A.M. Saeed. 2012. Effects of Gum Arabic ingestion on body mass index and body fat percentage in healthy adult females: two-arm randomized, placebo controlled, double-blind trial. Nutr. J. 11:111.:10.1186/1475-2891-1111-1111.

Baboota, R.K., M. Bishnoi, P. Ambalam, K.K. Kondepudi, S.M. Sarma, R.K. Boparai and K. Podili. 2013. Functional food ingredients for the management of obesity and associated co-morbidities—A review. Journal of Functional Foods. 5: 997–1012.

Baboota, R.K., N. Murtaza, S. Jagtap, D.P. Singh, A. Karmase, J. Kaur, K.K. Bhutani, R.K. Boparai, L.S. Premkumar, K.K. Kondepudi and M. Bishnoi. 2014a. Capsaicin-induced transcriptional changes in hypothalamus and alterations in gut microbial count in high fat diet fed mice. J. Nutr. Biochem. 25: 893–902.

Baboota, R.K., D.P. Singh, S.M. Sarma, J. Kaur, R. Sandhir, R.K. Boparai, K.K. Kondepudi and M. Bishnoi. 2014b. Capsaicin induces "brite" phenotype in differentiating 3T3-L1 preadipocytes. PLOS ONE. 9, e103093.

Bäckhed, F. 2014. Gut microbiota in metabolic syndrome. pp. 171–181 *In:* M. Orešič and A. Vidal-Puig (eds.). A Systems Biology Approach to Study Metabolic Syndrome. Springer.

Barneda, D., A. Frontini, S. Cinti and M. Christian. 2013. Dynamic changes in lipid droplet-associated proteins in the "browning" of white adipose tissues. Biochim. Biophys. Acta. 1831: 924–933.

Bashmakov, Y.K., S. Assaad-Khalil and I.M. Petyaev. 2011. Resveratrol may be beneficial in treatment of diabetic foot syndrome. Med. Hypotheses. 77: 364–367.

Bercik, P., S. Collins and E. Verdu. 2012. Microbes and the gut-brain axis. Neurogastroenterology & Motility. 24: 405–413.

Berni Canani, R., M. Di Costanzo and L. Leone. 2012. The epigenetic effects of butyrate: potential therapeutic implications for clinical practice. Clin. Epigenetics. 4: 4.

Bomhof, M.R., D.C. Saha, D.T. Reid, H.A. Paul and R.A. Reimer. 2014. Combined effects of oligofructose and Bifidobacterium animalis on gut microbiota and glycemia in obese rats. Obesity (Silver Spring). 22: 763–771.

Bondia-Pons, I., T. Hyötyläinen and M. Orešič. 2015. Role of microbiota in regulating host lipid metabolism and disease risk. pp. 235–260. In Metabonomics and Gut Microbiota in Nutrition and Disease. Springer.

Bonet, M.L., P. Oliver and A. Palou. 2013. Pharmacological and nutritional agents promoting browning of white adipose tissue. Biochim. Biophys. Acta. 1831: 969–985.

Brahe, L.K., A. Astrup and L.H. Larsen. 2013. Is butyrate the link between diet, intestinal microbiota and obesity-related metabolic diseases? Obes Rev. 14: 950–959.

Cani, P.D. and M. Van Hul. 2015. Novel opportunities for next-generation probiotics targeting metabolic syndrome. Cur. Opinin. Biotech. 32: 21–27.

Cani, P.D., C. Knauf, M.A. Iglesias, D.J. Drucker, N.M. Delzenne and R. Burcelin. 2006. Improvement of glucose tolerance and hepatic insulin sensitivity by oligofructose requires a functional glucagon-like peptide 1 receptor. Diabetes. 55: 1484–1490.

Cani, P.D., J. Amar, M.A. Iglesias, M. Poggi, C. Knauf, D. Bastelica, A.M. Neyrinck, F. Fava, K.M. Tuohy, C. Chabo, A. Waget, E. Delmee, B. Cousin, T. Sulpice, B. Chamontin, J. Ferrieres, J.F. Tanti, G.R. Gibson, L. Casteilla, N.M. Delzenne, M.C. Alessi and R. Burcelin. 2007. Metabolic endotoxemia initiates obesity and insulin resistance. Diabetes. 56: 1761–1772.

Cani, P.D., S. Possemiers, T. Van de Wiele, Y. Guiot, A. Everard, O. Rottier, L. Geurts, D. Naslain, A. Neyrinck, D.M. Lambert, G.G. Muccioli and N.M. Delzenne. 2009. Changes in gut microbiota control inflammation in obese mice through a mechanism involving GLP-2-driven improvement of gut permeability. Gut. 58: 1091–1103.

Cebra, J.J. 1999. Influences of microbiota on intestinal immune system development. Am. J. Clin. Nutr. 69: 1046s–1051s.

Charalampopoulos, D., R. Wang, S.S. Pandiella and C. Webb. 2002. Application of cereals and cereal components in functional foods: a review. Int. J. Food Microbio. 79: 131–141.

Chen, J., R. Wang, X.F. Li and R.L. Wang. 2012. Bifidobacterium adolescentis supplementation ameliorates visceral fat accumulation and insulin sensitivity in an experimental model of the metabolic syndrome. Br. J. Nutr. 107: 1429–1434.

Comalada, M., D. Camuesco, S. Sierra, I. Ballester, J. Xaus, J. Galvez and A. Zarzuelo. 2005. *In vivo* quercitrin anti-inflammatory effect involves release of quercetin, which inhibits inflammation through down-regulation of the NF-kappaB pathway. Eur. J. Immunol. 35: 584–592.

Consortium, H.M.P. 2012. Structure, function and diversity of the healthy human microbiome. Nature. 486: 207–214.

Cryan, J.F. and S. O'Mahony. 2011. The microbiome-gut-brain axis: from bowel to behavior. Neurogastroenterology & Motility. 23: 187–192.

Cummings, J.H. and A.M. Stephen. 2007. Carbohydrate terminology and classification. Eur. J. Clin. Nutr. 61: S5–S18.

de Luis, D.A., B. de la Fuente, O. Izaola, R. Aller, S. Gutierrez and M. Morillo. 2013. Double blind randomized clinical trial controlled by placebo with a fos enriched cookie on saciety and cardiovascular risk factors in obese patients. Nutr. Hosp. 28: 78–85.

Dehghan, P., B.P. Gargari, M.A. Jafar-Abadi and A. Aliasgharzadeh. 2014. Insulin controls inflammation and metabolic endotoxemia in women with type 2 diabetes mellitus: a randomized-controlled clinical trial. Int. J. Food Sci. Nutr. 65: 117–123.

Delzenne, N.M., A.M. Neyrinck, F. Backhed and P.D. Cani. 2011. Targeting gut microbiota in obesity: effects of prebiotics and probiotics. Nat. Rev. Endocrinol. 7: 639–646.

Dewulf, E.M., P.D. Cani, A.M. Neyrinck, S. Possemiers, A. Van Holle, G.G. Muccioli, L. Deldicque, L.B. Bindels, B.D. Pachikian, F.M. Sohet, E. Mignolet, M. Francaux, Y. Larondelle and N.M. Delzenne. 2011. Insulin-type fructans with prebiotic properties counteract GPR43 overexpression and PPARgamma-related adipogenesis in the white adipose tissue of high-fat diet-fed mice. J. Nutr. Biochem. 22: 712–722.

Dewulf, E.M., P.D. Cani, S.P. Claus, S. Fuentes, P.G. Puylaert, A.M. Neyrinck, L.B. Bindels, W.M. de Vos, G.R. Gibson, J.P. Thissen and N.M. Delzenne. 2013. Insight into the prebiotic concept: lessons from an exploratory, double blind intervention study with inulin-type fructans in obese women. Gut. 62: 1112–1121.

Dhiman, R.K., B. Rana, S. Agrawal, A. Garg, M. Chopra, K.K. Thumburu, A. Khattri, S. Malhotra, A. Duseja and Y.K. Chawla. 2014. Probiotic VSL#3 reduces liver disease severity and hospitalization in patients with cirrhosis: a randomized, controlled trial. Gastroenterology. 147: 1327–1337.e1323.

Duda-Chodak, A. 2012. The inhibitory effect of polyphenols on human gut microbiota. J. Physiol. Pharmacol. 63: 497–503.

El Khoury, D., C. Cuda, B.L. Luhovyy and G.H. Anderson. 2012. Beta glucan: health benefits in obesity and metabolic syndrome. J. Nutr. Metab. 2012: 851362.

Ellekilde, M., E. Selfjord, C.S. Larsen, M. Jakesevic, I. Rune, B. Tranberg, F.K. Vogensen, D.S. Nielsen, M.I. Bahl, T.R. Licht, A.K. Hansen and C.H.F. Hansen. 2014. Transfer of gut microbiota from lean and obese mice to antibiotic-treated mice. Sci. Rep. 4.

Eslamparast, T., H. Poustchi, F. Zamani, M. Sharafkhah, R. Malekzadeh and A. Hekmatdoost. 2014. Synbiotic supplementation in nonalcoholic fatty liver disease: a randomized, double-blind, placebo-controlled pilot study. Am. J. Clin. Nutr. 99: 535–542.

Everard, A., C. Belzer, L. Geurts, J.P. Ouwerkerk, C. Druart, L.B. Bindels, Y. Guiot, M. Derrien, G.G. Muccioli, N.M. Delzenne, W.M. de Vos and P.D. Cani. 2013. Cross-talk between *Akkermansia muciniphila* and intestinal epithelium controls diet-induced obesity. Proc. Natl. Acad. Sci. USA. 110: 9066–9071.

Fei, N. and L. Zhao. 2013. An opportunistic pathogen isolated from the gut of an obese human causes obesity in germfree mice. ISME J. 7: 880–884.

Fenzl, A. and F.W. Kiefer. 2014. Brown adipose tissue and thermogenesis. Horm. Mol. Biol. Clin. Investig. 19: 25–37.

Frost, G., M.L. Sleeth, M. Sahuri-Arisoylu, B. Lizarbe, S. Cerdan, L. Brody, J. Anastasovska, S. Ghourab, M. Hankir, S. Zhang, D. Carling, J.R. Swann, G. Gibson, A. Viardot, D. Morrison, E. Louise Thomas and J.D. Bell. 2014. The short-chain fatty acid acetate reduces appetite via a central homeostatic mechanism. Nat. Commun. 5: 3611.

Gallaher, D.D. and J. Khil. 1999. The effect of synbiotics on colon carcinogenesis in rats. The Journal of Nutrition. 129: 1483S–1487s.

Garcia, A.L., J. Steiniger, S.C. Reich, M.O. Weickert, I. Harsch, A. Machowetz, M. Mohlig, J. Spranger, N.N. Rudovich, F. Meuser, J. Doerfer, N. Katz, M. Speth, H.J. Zunft, A.H. Pfeiffer, and C. Koebnick. 2006. Arabinoxylan fibre consumption improved glucose metabolism, but did not affect serum adipokines in subjects with impaired glucose tolerance. Horm Metab. Res. 38: 761–766.

Guérin-Deremaux, L., M. Pochat, C. Reifer, D. Wils, S. Cho and L.E. Miller. 2011. The soluble fiber NUTRIOSE induces a dose-dependent beneficial impact on satiety over time in humans. Nutr. Res. 31: 665–672.

Herzig, S. and C. Wolfrum. 2013. Brown and white fat: from signaling to disease. Biochim. Biophys. Acta. 1831: 895.

Hidalgo, M., M.J. Oruna-Concha, S. Kolida, G.E. Walton, S. Kallithraka, J.P.E. Spencer, G.R. Gibson and S. de Pascual-Teresa. 2012. Metabolism of anthocyanins by human gut microflora and their influence on gut bacterial growth. J. Agric. Food Chem. 60: 3882–3890.

Joo, J.K. and K.S. Lee. 2014. Pharmacotherapy for obesity. J. Menopausal. Med. 20: 90–96.

Joyce, S.A., J. MacSharry, P.G. Casey, M. Kinsella, E.F. Murphy, F. Shanahan, C. Hill and C.G. Gahan. 2014. Regulation of host weight gain and lipid metabolism by bacterial bile acid modification in the gut. Proc. Natl. Acad. Sci. USA. 111: 7421–7426.

Kadooka, Y., M. Sato, K. Imaizumi, A. Ogawa, K. Ikuyama, Y. Akai, M. Okano, M. Kagoshima and T. Tsuchida. 2010. Regulation of abdominal adiposity by probiotics (*Lactobacillus gasseri* SBT2055) in adults with obese tendencies in a randomized controlled trial. Eur. J. Clin. Nutr. 64: 636–643.

Kadooka, Y., A. Ogawa, K. Ikuyama and M. Sato. 2011. The probiotic *Lactobacillus gasseri* SBT2055 inhibits enlargement of visceral adipocytes and upregulation of serum soluble adhesion molecule (sICAM-1) in rats. Int. Dairy Journal. 21: 623–627.

Kang, J.H., S.I. Yun, M.H. Park, J.H. Park, S.Y. Jeong and H.O. Park. 2013. Anti-obesity effect of *Lactobacillus gasseri* BNR17 in high-sucrose diet-induced obese mice. PLoS One. 8: e54617.

Kawabata, K., Y. Sugiyama, T. Sakano and H. Ohigashi. 2013. Flavonols enhanced production of anti-inflammatory substance(s) by Bifidobacterium adolescentis: prebiotic actions of galangin, quercetin, and fisetin. Biofactors. 39: 422–429.

Keenan, M.J., M. Janes, J. Robert, R.J. Martin, A.M. Raggio, K.L. McCutcheon, C. Pelkman, R. Tulley, M. Goita, H.A. Durham, J. Zhou and R.N. Senevirathne. 2013. Resistant starch from high amylose maize (HAM-RS2) reduces body fat and increases gut bacteria in ovariectomized (OVX) rats. Obesity (Silver Spring). 21: 981–984.

Kim, K.-A., J.-J. Jeong and D.-H. Kim. 2015. *Lactobacillus brevis* OK56 ameliorates high-fat diet-induced obesity in mice by inhibiting NF-κB activation and gut microbial LPS production. J. Funct. Foods. 13: 183–191.

Kolida, S. and G.R. Gibson. 2011. Synbiotics in health and disease. Annual Review of Food Science and Technology. 2: 373–393.

Kondepudi, K.K., P. Ambalam, I. Nilsson, T. Wadstrom and A. Ljungh. 2012. Prebiotic-non-digestible oligosaccharides preference of probiotic bifidobacteria and antimicrobial activity against *Clostridium difficile*. Anaerobe. 18: 489–497.

Kondepudi, K.K., M. Bishnoi, K. Podili, P. Ambalam, K. Mazumder, N. Murtaza, R.K. Baboota and R.K. Boparai. 2014. Dietary polysaccharides for the modulation of obesity via beneficial gut microbial manipulation. pp. 367–384. *In:* B. Noureddine (ed.). Polysaccharides: Natural Fibers in Food and Nutrition. CRC Press, Boca Raton, FL, USA.

Laparra, J.M. and Y. Sanz. 2010. Interactions of gut microbiota with functional food components and nutraceuticals. Pharmacol. Res. 61: 219–225.

Larrosa, M., M.J. Yañéz-Gascón, M.V. Selma, A. González-Sarrías, S. Toti, J.J. Cerón, F. Tomás-Barberán, P. Dolara and J.C. Espín. 2009. Effect of a low dose of dietary resveratrol on colon microbiota, inflammation and tissue damage in a DSS-induced colitis rat model. J. Agric. Food Chem. 57: 2211–2220.

Lebeer, S., J. Vanderleyden and S.C. De Keersmaecker. 2010. Host interactions of probiotic bacterial surface molecules: comparison with commensals and pathogens. Nat. Rev. Microbiol. 8: 171–184.

Lee, H.C., A.M. Jenner, C.S. Low and Y.K. Lee. 2006a. Effect of tea phenolics and their aromatic fecal bacterial metabolites on intestinal microbiota. Res. Microbiol. 157: 876–884.

Lee, H.Y., J.H. Park, S.H. Seok, M.W. Baek, D.J. Kim, K.E. Lee, K.S. Paek and Y. Lee. 2006b. Human originated bacteria, *Lactobacillus rhamnosus* PL60, produce conjugated linoleic acid and show anti-obesity effects in diet-induced obese mice. Biochim. Biophys. Acta. 1761: 736–744.

Lee, B.-H., Y.-H. Lo and T.-M. Pan. 2013. Anti-obesity activity of Lactobacillus fermented soy milk products. J. Funct. Foods. 5: 905–913.

Lin, H.V., A. Frassetto, E.J. Kowalik, Jr., A.R. Nawrocki, M.M. Lu, J.R. Kosinski, J.A. Hubert, D. Szeto, X. Yao and G. Forrest. 2012. Butyrate and propionate protect against diet-induced obesity and regulate gut hormones via free fatty acid receptor 3-independent mechanisms. PLoS ONE. 7: e35240.

Ma, K., P.K. Saha, L. Chan and D.D. Moore. 2006. Farnesoid X receptor is essential for normal glucose homeostasis. Journal of Clinical Investigation. 116: 1102.

Malaguarnera, M., M. Vacante, T. Antic, M. Giordano, G. Chisari, R. Acquaviva, S. Mastrojeni, G. Malaguarnera, A. Mistretta, G. Li Volti and F. Galvano. 2012. Bifidobacterium longum with fructo-oligosaccharides in patients with non alcoholic steatohepatitis. Dig. Dis. Sci. 57: 545–553.

Mazmanian, S.K., C.H. Liu, A.O. Tzianabos and D.L. Kasper. 2005. An immunomodulatory molecule of symbiotic bacteria directs maturation of the host immune system. Cell. 122: 107–118.

Meijer, K., P. de Vos and M.G. Priebe. 2010. Butyrate and other short-chain fatty acids as modulators of immunity: what relevance for health? Curr. Opin. in Clin. Nutr. & Metab. Care. 13: 715–721.

Meile, L., G. Le Blay and A. Thierry. 2008. Safety assessment of dairy microorganisms: Propionibacterium and Bifidobacterium. Int. J. Food Microbiol. 126: 316–320.

Moran, C.P. and F. Shanahan. 2014. Gut microbiota and obesity: Role in aetiology and potential therapeutic target. Best Pract & Res. Clin. Gastroenter. 28: 585–597.

Murtaza, N., R.K. Baboota, S. Jagtap, D.P. Singh, P. Khare, S.M. Sarma, K. Podili, S. Alagesan, T.S. Chandra, K.K. Bhutani, R.K. Boparai, M. Bishnoi and K.K. Kondepudi. 2014. Finger millet bran supplementation alleviates obesity-induced oxidative stress, inflammation and gut microbial derangements in high-fat diet-fed mice. Br. J. Nutr. 19: 1–12.

Naïma, S., R. Villeger, T.-S. Ouk, C. Delattre, M. Urdaci and P. Bressollier. 2014. Probiotics, prebiotics, and synbiotics for gut health benefits. Beneficial Microbes in Fermented and Functional Foods: 363.

Nathan, P.J., B.V. O'Neill, A. Napolitano and E.T. Bullmore. 2011. Neuropsychiatric adverse effects of centrally acting antiobesity drugs. CNS Neurosci. Ther. 17: 490–505.

Neish, A.S. 2009. Microbes in gastrointestinal health and disease. Gastroenterology. 136: 65–80.

Neyrinck, A.M., S. Possemiers, C. Druart, T. Van de Wiele, F. De Backer, P.D. Cani, Y. Larondelle and N.M. Delzenne. 2011. Prebiotic effects of wheat arabinoxylan related to the increase in bifidobacteria, Roseburia and Bacteroides/Prevotella in diet-induced obese mice. PLoS One. 6: e20944.

Neyrinck, A.M., V.F. Van Hee, N. Piront, F. De Backer, O. Toussaint, P.D. Cani and N.M. Delzenne. 2012. Wheat-derived arabinoxylan oligosaccharides with prebiotic effect increase satietogenic gut peptides and reduce metabolic endotoxemia in diet-induced obese mice. Nutr. Diabetes. 2: e28.:10.1038/nutd.2011.1024.

Ng, M., T. Fleming, M. Robinson, B. Thomson, N. Graetz, C. Margono, E.C. Mullany, S. Biryukov, C. Abbafati, S.F. Abera, J.P. Abraham, N.M.E. Abu-Rmeileh, T. Achoki, F.S. Al Buhairan, Z.A. Alemu, R. Alfonso, M.K. Ali, R. Ali, N.A. Guzman, W. Ammar, P. Anwari, A. Banerjee, S. Barquera, S. Basu, D.A. Bennett, Z. Bhutta, J. Blore, N. Cabral, I. C. Nonato, J.-C. Chang, R. Chowdhury, K.J. Courville, M.H. Criqui, D.K. Cundiff, K.C. Dabhadkar, L. Dandona, A. Davis, A. Dayama, S.D. Dharmaratne, E.L. Ding, A.M. Durrani, A. Esteghamati, F. Farzadfar, D.F.J. Fay, V.L. Feigin, A. Flaxman, M.H. Forouzanfar, A. Goto, M.A. Green, R. Gupta, N. Hafezi-Nejad, G.J. Hankey, H.C. Harewood, R. Havmoeller, S. Hay, L. Hernandez, A. Husseini, B.T. Idrisov, N. Ikeda, F. Islami, E. Jahangir, S.K. Jassal, S.H. Jee, M. Jeffreys, J.B. Jonas, E.K. Kabagambe, S.E.A.H. Khalifa, A.P. Kengne, Y.S. Khader, Y.-H. Khang, D. Kim, R.W. Kimokoti, J.M. Kinge, Y. Kokubo, S. Kosen, G. Kwan, T. Lai, M. Leinsalu, Y. Li, X. Liang, S. Liu, G. Logroscino, P.A. Lotufo, Y. Lu, J. Ma, N.K. Mainoo, G.A. Mensah, T.R. Merriman, A.H. Mokdad, J. Moschandreas, M. Naghavi, A. Naheed, D. Nand, K.M.V. Narayan, E.L. Nelson, M.L. Neuhouser, M.I. Nisar, T. Ohkubo, S.O. Oti, A. Pedroza, D. Prabhakaran, N. Roy, U. Sampson, H. Seo, S.G. Sepanlou, K. Shibuya, R. Shiri, I. Shiue, G.M. Singh, J.A. Singh, V. Skirbekk, N.J.C. Stapelberg, L. Sturua, B.L. Sykes, M. Tobias, B.X. Tran, L. Trasande, H. Toyoshima, S. van de Vijver, T.J. Vasankari, J.L. Veerman, G. Velasquez-Melendez, V.V. Vlassov, S.E. Vollset, T. Vos, C. Wang, X. Wang, E. Weiderpass, A. Werdecker, J.L. Wright, Y.C. Yang, H. Yatsuya, J. Yoon, S.-J. Yoon, Y. Zhao, M. Zhou, S. Zhu, A.D. Lopez, C.J.L. Murray and E. Gakidou. 2014. Global, regional, and national prevalence of overweight and obesity in children and adults during 1980–2013: a systematic analysis for the Global Burden of Disease Study 2013. The Lancet. 384: 766–781.

Nicholson, J.K., E. Holmes, J. Kinross, R. Burcelin, G. Gibson, W. Jia and S. Pettersson. 2012. Host-gut microbiota metabolic interactions. Science. 336: 1262–1267.

Niemelä, S., S. Miettinen, J. Sarkanen and N. Ashammakhi. 2008. Adipose tissue and adipocyte differentiation: molecular and cellular aspects and tissue engineering applications. Topics in Tissue Engineering. 4: 26.

Nilsson, A., E. Johansson, L. Ekstrom and I. Bjorck. 2013. Effects of a brown beans evening meal on metabolic risk markers and appetite regulating hormones at a subsequent standardized breakfast: a randomized cross-over study. PLoS One. 8: e59985.

Nunez, I.N., C.M. Galdeano, M. de LeBlanc Ade and G. Perdigon. 2014. Evaluation of immune response, microbiota, and blood markers after probiotic bacteria administration in obese mice induced by a high-fat diet. Nutrition. 30: 1423–1432.

O'Hara, A.M. and F. Shanahan. 2006. The gut flora as a forgotten organ. EMBO reports 7: 688–693.

Ogden, C.L., M.D. Carroll, B.K. Kit and K.M. Flegal. 2014. Prevalence of childhood and adult obesity in the united states, 2011–2012. JAMA. 311: 806–814.

Ohno, H. 2015. Impact of commensal microbiota on the host pathophysiology: focusing on immunity and inflammation. pp. 1–3. *In*: Seminars in immunopathology. Springer, Berlin, Heidelberg.

Okazaki, Y., N.V. Sitanggang, S. Sato, N. Ohnishi, J. Inoue, T. Iguchi, T. Watanabe, H. Tomotake, K. Harada and N. Kato. 2013. Burdock fermented by *Aspergillus awamori* elevates cecal Bifidobacterium, and reduces fecal deoxycholic acid and adipose tissue weight in rats fed a high-fat diet. Biosci. Biotechnol. Biochem. 77: 53–57.

Oso, A., A. Awe, F. Awosoga, F. Bello, T. Akinfenwa and E. Ogunremi. 2013. Effect of ginger (Zingiber officinale Roscoe) on growth performance, nutrient digestibility, serum metabolites, gut morphology, and microflora of growing guinea fowl. Tropical Animal Health and Production. 45: 1763–1769.

Ouchi, N., J.L. Parker, J.J. Lugus and K. Walsh. 2011. Adipokines in inflammation and metabolic disease. Nat. Rev. Immuno. 11: 85–97.

Paoli, A., G. Bosco, E. Camporesi and D. Mangar. 2015. Ketosis, ketogenic diet and food intake control: a complex relationship. Front. in Psycho. 6: 27.

Park, D.Y., Y.T. Ahn, C.S. Huh, S.M. Jeon and M.S. Choi. 2011. The inhibitory effect of *Lactobacillus plantarum* KY1032 cell extract on the adipogenesis of 3T3-L1 Cells. J. Med. Food. 14: 670–675.

Paturi, G., C.A. Butts, K.L. Bentley-Hewitt, D. Hedderley, H. Stoklosinski and J. Ansell. 2015. Differential effects of probiotics, prebiotics, and synbiotics on gut microbiota and gene expression in rats. J. Funct. Foods. 13: 204–213.

Poulos, S.P., M.V. Dodson and G.J. Hausman. 2010. Cell line models for differentiation: preadipocytes and adipocytes. Exp. Biol. Med. (Maywood). 235: 1185–1193.

Pouwels, P.H., R.J. Leer and W.J. Boersma. 1996. The potential of Lactobacillus as a carrier for oral immunization: development and preliminary characterization of vector systems for targeted delivery of antigens. J. of Biotech. 44: 183–192.

Pyra, K.A., D.C. Saha and R.A. Reimer. 2012. Prebiotic fiber increases hepatic acetyl CoA carboxylase phosphorylation and suppresses glucose-dependent insulinotropic polypeptide secretion more effectively when used with metformin in obese rats. J. Nutr. 142: 213–220.

Rakoff-Nahoum, S., J. Paglino, F. Eslami-Varzaneh, S. Edberg and R. Medzhitov. 2004. Recognition of commensal microflora by toll-like receptors is required for intestinal homeostasis. Cell. 118: 229–241.

Raman, M., I. Ahmed, P.M. Gillevet, C.S. Probert, N.M. Ratcliffe, S. Smith, R. Greenwood, M. Sikaroodi, V. Lam and P. Crotty. 2013. Fecal microbiome and volatile organic compound metabolome in obese humans with nonalcoholic fatty liver disease. Clin. Gastroentero. and Hepato. 11: 868–875. e863.

Reddy, B.S. and K. Watanabe. 1979. Effect of cholesterol metabolites and promoting effect of lithocholic acid in colon carcinogenesis in germ-free and conventional F344 rats. Cancer Res. 39: 1521–1524.

Rosen, B.S., K.S. Cook, J. Yaglom, D.L. Groves, J.E. Volanakis, D. Damm, T. White and B.M. Spiegelman. 1989. Adipsin and complement factor D activity: an immune-related defect in obesity. Science. 244: 1483–1487.

Salazar, N., E.M. Dewulf, A.M. Neyrinck, L.B. Bindels, P.D. Cani, J. Mahillon, W.M. de Vos, J.P. Thissen, M. Gueimonde, C.G. de Los Reyes-Gavilan and N.M. Delzenne. 2014. Inulin-type fructans modulate intestinal Bifidobacterium species populations and decrease fecal short-chain fatty acids in obese women. Clin. Nutr. 11: 00159–00159.

Sanders, M.E. 2008. Probiotics: definition, sources, selection, and uses. Clin. Infect. Dis. 46: S58–S61.

Sato, M., K. Uzu, T. Yoshida, E.M. Hamad, H. Kawakami, H. Matsuyama, I.A. Abd El-Gawad and K. Imaizumi. 2008. Effects of milk fermented by *Lactobacillus gasseri* SBT2055 on adipocyte size in rats. Br. J. Nutr. 99: 1013–1017.

Sayin, S.I., A. Wahlström, J. Felin, S. Jäntti, H.-U. Marschall, K. Bamberg, B. Angelin, T. Hyötyläinen, M. Orešič and F. Bäckhed. 2013. Gut microbiota regulates bile acid metabolism by reducing the levels of tauro-beta-muricholic acid, a naturally occurring FXR antagonist. Cell Metab. 17: 225–235.

Segata, N., D. Boernigen, T.L. Tickle, X.C. Morgan, W.S. Garrett and C. Huttenhower. 2013. Computational meta'omics for microbial community studies. Mol. Syst. Biology. 9: 666.

Sethi, J.K. and A.J. Vidal-Puig. 2007. Thematic review series: adipocyte biology. Adipose tissue function and plasticity orchestrate nutritional adaptation. J. Lip. Res. 48: 1253–1262.

Slavin, J. 2013. Fiber and prebiotics: mechanisms and health benefits. Nutrients. 5: 1417–1435.

Stojsavljevic, S., M. Gomercic Palcic, L. Virovic Jukic, L. Smircic Duvnjak and M. Duvnjak. 2014. Adipokines and proinflammatory cytokines, the key mediators in the pathogenesis of nonalcoholic fatty liver disease. World J. Gastroenterol. 20: 18070–18091.

Swann, J.R., E.J. Want, F.M. Geier, K. Spagou, I.D. Wilson, J.E. Sidaway, J.K. Nicholson and E. Holmes. 2011. Systemic gut microbial modulation of bile acid metabolism in host tissue compartments. Proc. Natl. Acad. Sci. USA. 108: 4523–4530.

Takemura, N., T. Okubo and K. Sonoyama. 2010. *Lactobacillus plantarum* strain No. 14 reduces adipocyte size in mice fed high-fat diet. Exp. Biol. Med. (Maywood). 235: 849–856.

Tang, W.W., Z. Wang, B.S. Levison, R.A. Koeth, E.B. Britt, X. Fu, Y. Wu and S.L. Hazen. 2013. Intestinal microbial metabolism of phosphatidylcholine and cardiovascular risk. N. Eng. J. Med. 368: 1575–1584.

Tetens, I. 2009. Scientific Opinion on the substantiation of health claims related to isomalto-oligosaccharides and maintenance of normal blood cholesterol concentrations (ID 817) pursuant to Article 13 (1) of Regulation (EC) No. 1924/2006: EFSA-Q-2008-1604. European Food Safety Authority.

Thomas, C., A. Gioiello, L. Noriega, A. Strehle, J. Oury, G. Rizzo, A. Macchiarulo, H. Yamamoto, C. Mataki and M. Pruzanski. 2009. TGR5-mediated bile acid sensing controls glucose homeostasis. Cell Metab. 10: 167–177.

Tilg, H. and A. Kaser. 2011. Gut microbiome, obesity, and metabolic dysfunction. J. Clin. Invest. 121: 2126.

Tinkov, A., O. Nemereshina, E. Popova, V. Polyakova, V. Gritsenko and A. Nikonorov. 2014. Plantago maxima leaves extract inhibits adipogenic action of a high-fat diet in female Wistar rats. European J. Nutr. 53: 831–842.

Torres-Fuentes, C., H. Schellekens, T.G. Dinan and J.F. Cryan. 2015. A natural solution for obesity: Bioactives for the prevention and treatment of weight gain. A review. Nutr. Neurosci. 18: 49–65.

Tremaroli, V. and F. Bäckhed. 2012. Functional interactions between the gut microbiota and host metabolism. Nature. 489: 242–249.

Trujillo, J.M., W. Nuffer and S.L. Ellis. 2015. GLP-1 receptor agonists: a review of head-to-head clinical studies. Ther. Adv. Endocrinol. Metab. 6: 19–28.

Tzounis, X., J. Vulevic, G.G. Kuhnle, T. George, J. Leonczak, G.R. Gibson, C. Kwik-Uribe and J.P. Spencer. 2008. Flavanol monomer-induced changes to the human faecal microflora. Br. J. Nutr. 99: 782–792.

Uehara, M. 2013. Isoflavone metabolism and bone-sparing effects of daidzein-metabolites. J. Clin. Biochem. Nutr. 52: 193–201.

Van den Abbeele, P., P. Gérard, S. Rabot, A. Bruneau, S. El Aidy, M. Derrien, M. Kleerebezem, E.G. Zoetendal, H. Smidt and W. Verstraete. 2011. Arabinoxylans and inulin differentially modulate the mucosal and luminal gut microbiota and mucin-degradation in humanized rats. Environmental Microbiology. 13: 2667–2680.

Ventura, M., F. Turroni, F. Strati and D. van Sinderen. 2014. The gut microbiota in health. Ed. by Julian R. Marchesi, Page 136 The Human Microbiota and Microbiome. CAB International, Nosworthy Way, Wallingford, Oxfordshire, UK.

Vidrine, K., J. Ye, R.J. Martin, K.L. McCutcheon, A.M. Raggio, C. Pelkman, H.A. Durham, J. Zhou, R.N. Senevirathne, C. Williams, F. Greenway, J. Finley, Z. Gao, F. Goldsmith and M.J. Keenan. 2014. Resistant starch from high amylose maize (HAM-RS2) and dietary butyrate reduce abdominal fat by a different apparent mechanism. Obesity (Silver Spring). 22: 344–348.

Vissiennon, C., K. Nieber, O. Kelber and V. Butterweck. 2012. Route of administration determines the anxiolytic activity of the flavonols kaempferol, quercetin and myricetin—are they prodrugs? J. Nutr. Biochem. 23: 733–740.

Vo, T.-S. and S.-K. Kim. 2013. Fucoidans as a natural bioactive ingredient for functional foods. J. Funct. Foods. 5: 16–27.

Vulevic, J., A. Juric, G. Tzortzis and G.R. Gibson. 2013. A mixture of trans-galactooligosaccharides reduces markers of metabolic syndrome and modulates the fecal microbiota and immune function of overweight adults. J. Nutr. 143: 324–331.

Wang, J., H. Tang, C. Zhang, Y. Zhao, M. Derrien, E. Rocher, J.E. van-Hylckama Vlieg, K. Strissel, L. Zhao, M. Obin and J. Shen. 2015. Modulation of gut microbiota during probiotic-mediated attenuation of metabolic syndrome in high fat diet-fed mice. ISME J. 9: 1–15.

Watanabe, M., S.M. Houten, C. Mataki, M.A. Christoffolete, B.W. Kim, H. Sato, N. Messaddeq, J.W. Harney, O. Ezaki and T. Kodama. 2006. Bile acids induce energy expenditure by promoting intracellular thyroid hormone activation. Nature. 439: 484–489.

Weng, H., K. Endo, J. Li, N. Kito and N. Iwai. 2015. Induction of Peroxisomes by Butyrate-Producing Probiotics. PLoS ONE 10: e0117851.

Whittle, A., J. Relat-Pardo and A. Vidal-Puig. 2013. Pharmacological strategies for targeting BAT thermogenesis. Trends Pharmacol. Sci. 34: 347–355.

WHO. 2012. World Health Statistics 2012-Indicator compendium. World Health Organization. http://www.who.int/gho/publications/world_health_statistics/WHS2012_IndicatorCompendium.pdf (accessed on 23/02/2015).

WHO. 2014a. Childhood overweight and obesity. Global Strategy on Diet, Physical Activity and Health. World Health Organization, Geneva. http://www.who.int/dietphysicalactivity/childhood/en/(accessed on 23/02/2015).

WHO. 2014b. Obesity and overweight. World Health Organization. http://www.who.int/mediacentre/factsheets/fs311/en/(accessed on 23/02/2015).

Wong, J.M., R. de Souza, C.W. Kendall, A. Emam and D.J. Jenkins. 2006. Colonic health: fermentation and short chain fatty acids. J. Clin. Gastroentero. 40: 235–243.

Xiao, S., N. Fei, X. Pang, J. Shen, L. Wang, B. Zhang, M. Zhang, X. Zhang, C. Zhang, M. Li, L. Sun, Z. Xue, J. Wang, J. Feng, F. Yan, N. Zhao, J. Liu, W. Long and L. Zhao. 2014. A gut microbiota-targeted dietary intervention for amelioration of chronic inflammation underlying metabolic syndrome. FEMS Microbiol. Ecol. 87: 357–367.

Zarrati, M., E. Salehi, K. Nourijelyani, V. Mofid, M.J. Zadeh, F. Najafi, Z. Ghaflati, K. Bidad, M. Chamari, M. Karimi and F. Shidfar. 2014. Effects of probiotic yogurt on fat distribution and gene expression of proinflammatory factors in peripheral blood mononuclear cells in overweight and obese people with or without weight-loss diet. J. Am. Coll. Nutr. 33: 417–425.

Zhang, Z., X. Peng, S. Li, N. Zhang, Y. Wang and H. Wei. 2014. Isolation and identification of quercetin degrading bacteria from human fecal microbes. PLoS One. 9: e90531.

Index